Computational Linguistics and Talking Robots

Roland Hausser

Computational Linguistics and Talking Robots

Processing Content in Database Semantics

 Springer

Prof. Dr. Roland Hausser
Friedrich-Alexander-Universität
Erlangen-Nürnberg
Abteilung für Computerlinguistik
Bismarckstraße 12
91054 Erlangen
Germany
rrh@linguistik.uni-erlangen.de

ISBN 978-3-642-42998-9   ISBN 978-3-642-22432-4 (eBook)
DOI 10.1007/978-3-642-22432-4
Springer Heidelberg Dordrecht London New York

ACM Codes: I.2.7, H.5.2, I.2.9, I.2.1
© Springer-Verlag Berlin Heidelberg 2011
Softcover re-print of the Hardcover 1st edition 2011

Cover photo: Sphinx by Jakob C. Schletterer (1699–1774) Kloster Altenburg, 1735 Niederösterreich
        © Julia Liekenbrock 2011

Cover design: WMXDesign GmbH, Heidelberg

Printed on acid-free paper

Springer is part of Springer Science+Business Media (www.springer.com)

Preface

The practical task of building a talking robot requires a theory of how natural language communication works. Conversely, the best way to computationally verify a theory of free natural language communication is by demonstrating its functioning concretely in the form of a talking robot as the epitome of human-machine communication.

To build an actual robot requires hardware which provides recognition and action *interfaces* appropriate for this task – like those of **C3PO**, the bronzy protocol robot in the **Star Wars** movies. Because such a hardware is presently hard to come by, our method is theoretical: we present an artificial cognitive agent with language as a software system called Database Semantics (DBS).

DBS complements work in robotics which does include the interface hardware for reconstructing the sensorimotor loop. For example, Roy (2003, 2008) has been working on an actual robot with language, called **Ripley**.[1]

Ripley and the DBS system have in common that they are *grounded*. This is because both approaches are agent-oriented, in contradistinction to Phrase Structure Grammar and Truth-Conditional Semantics, which are sign-oriented.

Unlike the fictional C3PO, the real Ripley consists of a robotic arm mounted on a table.[2] Attached to the arm is a hand, a video camera, a microphone, and a loudspeaker. To manage the difficulties of building a robot as a hardware machine, Ripley is focused "on *simple* language use, such as that of small children" (Roy 2008, p. 3). The work concentrates on implementing examples, such as "Human: **Hand me the blue one**," referring to a cup on Ripley's table (op. cit., p. 10). This approach requires expending much time and labor on practical troubles like motor burnout from overexertion (op. cit., p. 19).

A theoretical approach, in contrast, does not have to deal with the technical difficulties of hardware engineering. Thus there is no reason to simplify our system to a toddler's use of language. Instead the software components of DBS aim at completeness of function and data coverage in word form recognition, syntactic-semantic interpretation, inferencing, and so on, leaving the procedural implementation of elementary concepts, e.g., *blue* or *cup*, for later.

---

[1] Thanks for a friendly welcome at the MIT Media Lab during a visit in 2002.

[2] For a similar approach see Dominey and Warneken (in press).

In the meantime, DBS uses placeholders for concepts assumed to be implemented as elementary recognition and action procedures of the agent. Embedded into flat feature structures called proplets, the concepts help to build the agent's *context component* for non-language cognition, and are reused as meanings of the agent's *language component*. In the hear and speak modes, the two components interact automatically on the basis of *pattern matching*.

Language proplets differ from context proplets in that language proplets have a language-dependent surface[3] as value, while context proplets do not. As a consequence, the context level may be constructed from the language level simply by omitting the surfaces of the language proplets (4.3.3).

The export of language-level constructs to the context level allows us to make the structural corelation between the two levels simple and direct. For example, there is no processing in language interpretation and production except for the literal mapping from the context to the language component in the speak mode and from the language to the context component in the hear mode. All non-literal interpretation, for example, of a metaphor, is done by inferencing before the utterance in the speak mode and after in the hear mode.[4]

The compositional semantics of the language component, i.e., coordination and functor-argument, intra- and extrapropositional, and the basic parts of speech, are reused in the context component. By connecting the context component to the agent's interfaces, recognition and action at the non-language level may be processed in the same way as at the language level.

The direct interaction between the language and the context[5] components implies linguistic relativism, also known as the Humboldt-Sapir-Whorf hypothesis. It is counterbalanced by the hypothesis that the different natural languages all work essentially the same way. This is supported by the wide range of languages[6] which have been analyzed using the DBS approach.

The similarity of language processing and context processing in DBS is motivated by function and simplicity of design. The similarity poses also an empirical question in psychology and neurology. It is not implausible, however, that language processing and context processing are cut from the same cloth.

---

[3] For example, dog, chien, Hund, and cane are different surfaces for the same concept (6.6.3).

[4] There are two kinds of inferencing in DBS, language inferencing and content inferencing. Both have the same form and work the same way, but language inferencing computes relations coded into the surface, while content inferencing models reasoning which is independent of language. Both kinds of inferencing contribute equally to language interpretation and production.

[5] Of course, there are areas of non-language cognition, e.g., reflexes (Sect. 6.1), which do not have counterparts in language cognition.

[6] See p. 46 for a list of natural languages which DBS has been applied to so far. Language-specific lexicalization and syntactic-semantic constructions require custom work and contribute to an ever growing library of meanings and constructions in different natural languages.

Acknowledgments

Thanks to Brian MacWhinney, Department of Psychology, Carnegie Mellon University, Pittsburgh, USA; Haitao Liu, Institute of Applied Linguistics, Zhejiang University, Hangzhou, China; Minhaeng Lee, Department of German, Yonsei University, Seoul, Korea; and Kiyong Lee, Department of Linguistics, Korea University, Seoul, Korea, for their most helpful comments on prefinal versions of the manuscript. Thanks to the members of my team at the CLUE (Computational Linguistics U. Erlangen), Johannes Handl, Besim Kabashi, Thomas Proisl, and Carsten Weber, who helped in the development and the proof-reading of the text at every step of the way. Thanks also to Andreas Struller, who helped with the LaTeX. All remaining mistakes are mine.

May 2011   Roland Hausser
Friedrich-Alexander
Universität
Erlangen-Nürnberg

Remark on Footnotes and References

The text contains many bibliographical references and footnotes. These are intended for readers who want to explore a topic in greater depth. The casual reader may just ignore the references and footnotes, and stick to the text alone.

Website for online papers and list of publications

http://www.linguistik.uni-erlangen.de/clue/en/publikationen.html

Abbreviations Referring to Previous Work

SCG'84, NEWCAT'86, CoL'89, TCS'92, FoCL'1999, AIJ'01, L&I'05, NLC'06, and L&I'10 are defined as follows:

SCG'84   Hausser, R. (1984) *Surface Compositional Grammar*, p. 274, München: Wilhelm Fink Verlag

NEWCAT'86   Hausser, R. (1986) *NEWCAT: Natural Language Parsing Using Left-Associative Grammar*. (Lecture Notes in Computer Science 231), p. 540, Springer

CoL'89   Hausser, R. (1989) *Computation of Language: An Essay on Syntax, Semantics and Pragmatics in Natural Man-Machine Communication*, Symbolic Computation: Artificial Intelligence, p. 425, Springer

TCS'92   Hausser, R. (1992) "Complexity in Left-Associative Grammar," *Theoretical Computer Science*, Vol. 106.2: 283–308, Amsterdam: Elsevier

FoCL'99   Hausser, R. (1999/2001) *Foundations of Computational Linguistics, Human-Computer Communication in Natural Language, 2nd ed. 2001*, p. 578, Springer

AIJ'01   Hausser, R. (2001) "Database Semantics for Natural Language." *Artificial Intelligence*, Vol. 130.1:27–74, Amsterdam: Elsevier

L&I'05   Hausser, R. (2005) "Memory-Based Pattern Completion in Database Semantics," *Language and Information*, Vol. 9.1: 69–92, Seoul: Korean Society for Language and Information

NLC'06   Hausser, R. (2006) *A Computational Model of Natural Language Communication – Interpretation, Inferencing, and Production in Database Semantics*, p. 365, Springer

L&I'10   Hausser, R. (2010) "Language Production Based on Autonomous Control – A Content-Addressable Memory for a Model of Cognition," *Language and Information*, Vol. 11: 5–31, Seoul: Korean Society for Language and Information

Contents

Part III. Final Chapter

1. Introduction: How to Build a Talking Robot

Can computational linguistics steer clear of artificial cognitive agents with language, i.e., talking robots? The answer is *no* if the research goal of our interdisciplinary field is a functional reconstruction of free natural language communication, verified by a computational model. The practical outcome of this approach is unrestricted human-machine communication in natural language.

A talking robot requires language cognition as well as nonlanguage cognition. For example, a human telling the artificial agent what to do requires the machine to understand natural language and to perform nonlanguage action. Similarly, for the artificial agent to tell a human what is going on requires the machine to have nonlanguage recognition and natural language production.[1]

The essential contributions of the robot metaphor to computational linguistics are the need (i) for a distinction between the robot-external environment and the robot-internal cognition, (ii) for interfaces providing a connection between the robot's environment and cognition, and (iii) for an autonomous control connecting the robot's recognition and action in a meaningful way.[2]

We honor these plain facts by making them founding assumptions of Database Semantics (DBS). They provide valuable requirements and restrictions which differ from the founding assumptions of Symbolic Logic in analytic philosophy and of Nativism in theoretical linguistics. They also provide excellent heuristics for designing the functional flow of the DBS software.

1.1 Universals

DBS defines successful language communication succinctly as a transfer of information from the cognition of the speaker to the cognition of the hearer, solely by means of a time-linear sequence of modality-dependent unanalyzed external language surfaces (Chap. 2). The transfer is successful if the information coded by the speaker is reconstructed equivalently by the hearer.

---

[1] Cf. FoCL'99, Sect. 23.5, for the ten SLIM states of cognition.

[2] There are additional requirements, such as an agent-internal database, but these are not robot-specific.

The theoretical and computational modeling of this transfer mechanism is based on the following properties, which are universal[3] among the natural languages of the world:

1.1.1 UNIVERSALS OF NATURAL LANGUAGE COMMUNICATION

1. The cycle of natural language communication is based on the *hear*, the *think*, and the *speak* modes of cognitive agents.
2. In communication, expressions of natural language are interpreted relative to an agent-internal *context* of use.
3. All natural languages have a *time-linear* structure, i.e., linear like time and in the direction of time.
4. All natural languages use the three kinds of sign *symbol, index,* and *name,* each with its own mechanism of reference.
5. All natural languages use *coordination* and *functor-argument*[4] to compose content at the *elementary*, the *phrasal*, and the *clausal* level.
6. All natural languages distinguish parts of speech, e.g., *noun* (object, argument), *verb* (relation, functor), and *adjective* (property, modifier).[5]
7. All natural languages have the sentential moods *declarative, interrogative,* and *imperative*.

The above universals characterize the agent-oriented approach of DBS, in contradistinction to the sign-oriented approaches of today's linguistics[6] and philosophy of language.

The language communication universals of DBS provide building blocks for and constraints on the design of a talking robot. Once the natural language

---

[3] The differences between natural languages are comparatively minor (cf. NLC'06, 4.6.1). In this respect, we agree with Chomsky. However, while Nativism formulates the similarity between languages as an innate Universal Grammar, DBS derives it from a common transfer mechanism.

Regarding Nativism from a history of science point of view, it is notable that this fragmented field has been sworn again and again to the position that the different schools really all want the same, e.g., Wasow, Postscript to Sells 1985. If they all want the same, why are there so many different schools? Perhaps the gravitational pull of the central theory is not sufficient, resulting naturally in a disintegration into all kinds of variants. This may be because the empirical base of a sign-oriented approach is simply not broad enough. As an agent-oriented approach, DBS broadens the empirical base substantially; it is also desigend from the outset to provide computational verification.

[4] We use functor-argument throughout for what is commonly called functor-argument structure, simply because it is shorter and more proportionate to its sister, coordination.

[5] Whether all natural languages have the same basic parts of speech is discussed controversially in language typology and difficult to decide (Sect. 3.6).

[6] See MacWhinney (2004) for an evaluation of the Nativist universals from the viewpoint of cognitive psychology.

communication mechanism has been implemented as a basic, high-level software machine, it may be loaded with different languages and different contents for different applications.

At the most general level, the DBS system requires the following constructs:

1.1.2 REQUIREMENTS OF A GROUNDED ARTIFICIAL AGENT

In order to be grounded, a cognitive agent requires a body with

1. *interfaces*
   for recognition and action, based on
2. a *data structure*
   for representing content,
3. a *database*
   for storing and retrieving content,
4. an *algorithm*
   for reading content into and out of the database as well as for processing content, and combined into
5. a *software program*
   which models the cycle of natural language communication as well as language and nonlanguage inferencing.

The interfaces for language as well as nonlanguage content must be suitable for connecting to the recognition and action hardware of future robots when they become available.

Until then, we have to make do with today's general-purpose computers. Despite their limitation to the keyboard for recognition and the screen for action as their main interfaces, they allow us to run the language interpretation, the language production, and the reasoning software. The computational implementation allows the researchers to observe the cognitive operations of the artificial agent directly (tracing), which is of great heuristic value.

During natural language communication, the DBS system requires the following *conversions* into which all upscaling must be embedded:

1.1.3 THE CONVERSION UNIVERSALS OF DBS

1. From the agent's *speak* mode to its *hear* mode and back (Chap. 3),
2. from a *modality-free* to a *modality-dependent* representation of the surface in the speak mode and back in the hear mode, in word form production (synthesis) and recognition (Chap. 2),

3. from *order-free content* to *ordered surfaces* in the speak mode and back in the hear mode (Chap. 3), and

4. from the STAR-0 to the STAR-1 perspective in the speak mode and from the STAR-1 to the STAR-2 perspective in the hear mode (Chaps. 10, 11).[7]

Given that each kind of conversion is realized in two directions, there are altogether eight. Thereby, the conversions 2–4 are embedded into conversion 1 between the *speak* mode and the *hear* mode, as shown graphically below:

1.1.4 INTERNAL STRUCTURE OF THE DBS CONVERSION UNIVERSALS

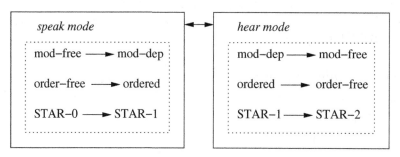

The Conversion Universals combine into a fundamental framework of structure and function, which forms the backbone of the DBS software machine.

1.2 Computational Verification

The continuous advancement of the overall system, by upscaling the lexicon and the rules for language and nonlanguage cognition, is an important practical goal for many applications in human-machine communication. It is also a method of computational verification for the overall DBS system and the SLIM theory of language (FoCL'99, NLC'06) on which it is built.

The long history of natural and cognitive science provides many examples of theories which sooner or later revealed some deep-seated flaw. In such a case, the theory cannot be fixed without completely uprooting its conceptual and, if present, mathematical framework, and in consequence much of its empirical work, which is unfortunate.

Nevertheless, discovering such flaws is necessary for the progress of science and constitutes a method of verification in the following sense: Theories shown to have a fatal flaw[8] are verified to be unsuitable for further upscaling, while

---

[7] A STAR represents the parameters of Space, Time, Agent, and Recipient. Cf. FoCL'99, Sect. 5.3.
[8] Sect. 2.6; FoCL'99, Sects. 8.5, 19.5.

theories capable of accommodating continuous upscaling at a nontrivial level are verified to be suitable for further development.

Therefore, the long-term success of upscaling the overall system must be a major concern in today's linguistics. For empirical support, an efficient implementation is a methodological necessity because without it the relative correctness of the theory cannot be tested automatically and objectively on large sets of relevant data.[9] After implementations in LISP and C, DBS is currently implemented as the JSLIM software written in Java.

Of course, we can only hope that the design of DBS will sustain the upscaling effort long-term. We are optimistic, however, because the cognitive distinctions and building blocks listed in 1.1.1 – 1.1.4 are so general, obvious, and simple that an alternative model of natural language communication doing without them is likely to have serious flaws.

1.3 Declarative Specification

For theoretical and practical reasons, we present the DBS theory of natural language communication as a *declarative specification* in the sense of computer science. In contradistinction to an *algebraic definition* in logic (which is based solely on set theory[10]), a declarative specification may describe a software machine in terms of such higher level notions as interfaces, components, functional flow, data structure, memory, and algorithms (cf. AIJ'01).

A declarative specification abstracts away from the *accidental* properties of an implementation (program), such as the choice of the hardware, the programming language, and the idiosyncrasies of the programmer. Instead, a declarative specification defines the *necessary* properties of the software and of the theory on which it is based. As a consequence, one declarative specification may have an open number of different implementations which are all equivalent with respect to the necessary properties.[11]

---

[9] This method of verification differs from those of the natural sciences (repeatability of experiments) and mathematics (proof of consistency), though it is compatible with them; cf. NLC'06, Sect. 1.2.

[10] Pure set theory is not equipped to represent the computationally relevant aspects of a software machine. High level modeling languages like UML (Universal Modeling Language, e.g., Fowler 2004) and ER (Entity Relationship, e.g., Chen 1976, Beynon-Davies 2003) developed in computer science are not flexible and detailed enough to address all the issues of DBS – though reconstructing certain aspects of the theory in UML, ER, or other projects of similar intent could well be worthwhile.

[11] The dichotomy between a declarative specification and its possible procedural implementations differs from the Nativist dichotomy between competence (language knowledge) and performance (processing knowledge). The latter distinction is used for "competence grammars" intended to model an innate human language ability, reified as a language acquisition device (LAD).

In this connection it is worth noting that Chomsky, Sag, and others have emphasized repeatedly

A crucial part of a declarative specification is the choice of the basic unit – analogous to that of other sciences founded on basic units such as the cell in biology or the atom in chemistry and physics. The basic unit of DBS is the *pro-plet*, defined as a flat feature structure[12] with a finite set of ordered attributes and double-ended queues as values. Compared to the recursive feature structures used in Nativism (with unordered attributes, but an order of embedding), proplets have the following advantages:

1.3.1 ADVANTAGES OF PROPLETS

1. Flat ordered feature structures are easier to read and computationally more efficient than recursive feature structures with unordered attributes.
2. Flat ordered feature structures provide for easy schema derivation and for easy pattern matching.
3. The combination of a proplet's core and **prn** value provides a natural primary key for storage in and retrieval from memory.
4. Coding the semantic relations between proplets as addresses (Sect. 4.4) makes proplets order-free and therefore amenable to the needs of one's database.
5. The semantic relations between proplets enable time-linear navigation along those relations, reintroducing order and serving as the selective activation of content, as needed in language production and inferencing.

In summary, a proplet is defined as a *list* of features (internal order) and each feature is defined as an attribute-value pair (**avp**). The proplets representing a complex content, in contrast, are a *set* (no external order, e.g., 3.2.1).

The data structure of proplets has been used in the LA-hear, LA-think, and LA-speak grammars defined in FoCL'99 and NLC'06. They map sequences of natural language surfaces into sets of proplets (hear mode), and sets of proplets into corresponding sequences of natural language surfaces (speak mode).

Proplets turn out to be versatile in that they maintain their format and their formal properties in a multitude of different functions. Consider the following examples of a language proplet for German, a language proplet for French, a content proplet, and a pattern proplet:

---

that competence grammars are *not intended* to model the language processing by the speaker-hearer. Therefore it is hardly surprising that the functional flow of competence grammars (NLC'06, 3.4.5) is incompatible with that of a talking robot. Also, competence grammars are cumbersome to program because they fail to be type-transparent (Berwick and Weinberg 1984; cf. FoCL'99, Sect. 9.3).

The declarative specification of DBS, in contrast, is an entirely computational theory. It aims at completeness of function and of data coverage as well as low complexity because the system must be efficient enough to run in real-time.

1.3.2 COMPARISON OF LANGUAGE, CONTENT, AND PATTERN PROPLETS

German proplet	*French proplet*	*content proplet*	*pattern proplet*
sur: Hund noun: dog cat: m-g sem: sg fnc: bark mod: old prn: 23	sur: chien noun: dog cat: sn m sem: sg fnc: bark mod: old prn: 24	sur: noun: dog cat: sn sem: sg fnc: bark mod: old prn: 25	sur: noun: α cat: sn sem: sg fnc: mod: prn: K

Proplets for a word in different languages differ mainly in their sur(face) value, here Hund vs. chien. Language and content proplets differ in the presence or absence of a sur value. Pattern and content proplets differ in the presence or absence of values defined as variables, here α and K. In addition there is the distinction between lexical proplets and connected proplets. The latter represent content resulting from hear mode derivations (e.g., 3.3.1), nonlanguage recognition (Chap. 8), and inferencing (Chaps. 5, 6, 10, 11).[13]

1.4 Time-Linearity

The development of DBS may be traced back to a Montague-style formal "fragment" of natural language, published as SCG'84.[14] The goal was to realize the First Principle of Pragmatics (PoP-1)[15] in a system of Truth-Conditional Semantics. This gave rise to the following question:

Is the agent part of the model, or is the model part of the agent?

For analyzing climate change, pollution, the stock market, etc., the first option is appropriate. For reconstructing cognition, the second option must be taken.

---

[12] A precursor of feature structures in linguistics is a list of binary values without attributes, called a "feature bundle", e.g., $\begin{bmatrix} +vocalic \\ +high \end{bmatrix}$, used by Chomsky and Halle (1968) for purposes of morphophonology. The introduction of feature structures with attribute-value pairs is credited to Minsky (1975). Called "frames" by Minsky, their origin may be traced back to the legendary Dartmouth workshop of 1956, attended also by Newell, Simon, McCarthy, Rochester, and Shannon.

[13] In the JSLIM software (Weber et al. 2010), proplets always have all their attributes in the appropriate order (cf. NLC'06, Sect. 13.1). In this text, however, proplets are handled in a more liberal fashion: they are often shown with only a subset of their attribute-value pairs. The purpose is a more succinct presentation and drawing attention to properties relevant to the issue at hand.

[14] Thanks to N. Belnap and R. Thomason at the Philosophy Department of the University of Pittsburgh (1978–1979), and to J. Moravscik and G. Kreisel at the Philosophy Department of Stanford University (1979–1980, 1983–1984) for revealing the secrets of a Tarskian semantics. Later, at Carnegie Mellon University (1986–1989), understanding this fundamental topic was helped further by Dana Scott. Cf. FoCL'99, Chaps. 19–21. The 1978–1979 stay in Pittsburgh and the 1979–1980 stay at Stanford were supported by a two-year DFG research grant.

[15] PoP-1: *The speaker's utterance meaning₂ is the use of the sign's literal meaning₁ relative to an internal context.* FoCL'99, 4.3.3.

This alternative is at the bottom of the distinction between a *sign-oriented* and an *agent-oriented* approach. Truth-Conditional Semantics is sign-oriented because it models meaning as a direct relation between a sign and a possible world, which includes the agent(s). SCG'84 and the later stages of DBS are agent-oriented because they treat the model as part of the agent.

Another problem with the Montague-style fragment of SCG'84 was Categorial Grammar. Designed by Leśniewski (1929) and Ajdukiewicz (1935), the combinatorics are coded into the lexical categories, using only two rules in a nondeterministic bottom-up derivation order (cf. FoCL'99, Sects. 7.4, 7.5). Categorial Grammar is elegant and simple intuitively, but the rule applications are based completely on trial and error. This is so inefficient computationally that it frustrated our attempt to program the SCG'84 fragment in LISP.[16]

The experience confirmed the more general insight that a rigorous formalization is only a necessary, but not a sufficient, condition for ensuring a reasonable implementation as a well-designed computer program. What is needed in addition is the declarative specification of the interfaces, the components, the functional flow, and the motor driving the derivation, as well as a general idea of how natural language communication works.

Our solution to the computational inefficiency of Categorial Grammar is a strictly time-linear derivation order. The algorithm of DBS, called LA-grammar, parses an input language surface one word form after the other, from beginning to end. This means theoretically and technically that the derivation order of LA-grammar is left-to-right (left-associative, time-linear), based on computing *possible continuations*. The hierarchical systems of Phrase Structure Grammar and Categorial Grammar, in contrast, compute *possible substitutions*[17] in a top-down or bottom-up derivation order.

LA-grammar was first implemented in LISP.[18] The program, published in NEWCAT'86, demonstrated the time-linear analysis of 221 constructions of German and 114 constructions of English, and was re-implemented by readers in South America, Switzerland, and other locations.[19] Later, an algebraic definition[20] was distilled from the NEWCAT'86 source code.

---

[16] Even today, no implementation of a Categorial Grammar exists for any fragment of a natural language with nontrivial data coverage. Such an implementation would be most useful for working with this formal theory. Without it, Categorial Grammars have found no direct practical applications in their long history.

[17] A notable exception is the time-linear system of Kempson et al. (2001). See also Lee (2002).

[18] Thanks to Stanley Peters and the CSLI for the 1984–1986 stay at Stanford University. By generously providing the then newest technology, the CSLI made programming the first left-associative parser possible, initially in Maclisp on a DEC Tops-20 mainframe, then in Interlisp-D on Xerox Star 8010 Dandelion and Dandetiger workstations, and finally in Common Lisp on an HP 9000 machine.

[19] These efforts became known because one little function had been omitted accidentally in the NEWCAT'86 source code publication, causing several of the re-programmers to write and ask for it.

1.5 Surface Compositionality

The design of LA-grammars for natural languages is guided by the methodological principle of Surface Compositionality (cf. FoCL'99, Sect. 4.5). It prohibits against tricks, such as the use of word forms with a meaning but an empty surface or with a surface but an empty meaning – all for the sake of empty "linguistic generalizations" (cf. FoCL'99, Sect. 4.4).[21]

By interpreting Surface Compositionality as the condition that each time-linear derivation step must "eat" a next word form in the input surface, the computation of possible continuations at any point in the derivation is limited to the small number of rules in the current rule package. The definition of (i) an upper bound on the complexity of an LA-rule application and (ii) different degrees of ambiguity resulted in the LA-grammar complexity hierarchy.

This hierarchy is the first, and so far the only, complexity hierarchy which is orthogonal to the Chomsky hierarchy of Phrase Structure Grammar (TCS'92). For example, the formal language $a^k b^k$ is polynomial (context-free) while $a^k b^k c^k$ is exponential (context-sensitive) in the Chomsky hierarchy (Aho and Ullman 1977), whereas in DBS they are both C1-languages and therefore of only linear complexity.

The LA-hierarchy of languages and complexity classes raised the question of where natural language is located. Given that the only source of complexity in the large class of C-languages are *recursive* ambiguities, we investigated whether or not this kind of ambiguity may be found in natural languages. A **yes** would mean a polynomial or exponential complexity of the natural languages. A **no** would mean that the natural languages parse in linear time.

The latter is what one would expect naturally – an argument elaborated by Gazdar (1982). So far, detailed syntactic-semantic research found that sequences of prepositional phrases like **on the table under the tree behind the house** are (i) the only candidates for a recursive ambiguity in natural language, but (ii) allow an alternative analysis which runs in linear time (Sect. 8.3; FoCL'99, Sects. 12.5, 21.5; NLC'06, Chap. 15).

---

[20] Thanks to Stuart Shieber and Dana Scott, who at different times and places helped in formulating the algebraic definition for LA-grammar, published in CoL'89 and TCS'92.

[21] Lately, HPSG has adopted some version of Surface Compositionality, called "surface-oriented" (Sag and Wasow 2011). 27 years after SCG'84, this is a substantial step in the right direction. The next step would be a metamorphosis into an agent-oriented approach. To be genuine, it would have to model the cycle of natural language communication, which requires a time-linear derivation order at least for the speak and the hear modes (Sect. 2.6). This, however, would require HPSG to part with much of its Nativist baggage.

Parallel to the work on complexity, the NEWCAT'86 parser was followed up by the CoL'89 parser, also implemented in LISP.[22] Its purpose was to provide the earlier parser with a semantic interpretation. Because the far-reaching ramifications of an agent-oriented approach had not yet become clear, the most promising goal seemed to show that a strictly time-linear, strictly surface compositional syntax can produce traditional semantic hierarchies.[23]

The CoL'89 parser provides syntactic-semantic analyses for 421 different grammatical constructions of English. The derivation of semantic hierarchies, based on a time-linear, surface compositional LA-grammar, was implemented using the FrameKit software (Carbonell and Joseph 1986). This worked well for producing the semantic trees for all constructions of the CoL'89 fragment.

Our next goal was to model (i) language interpretation as an embedding of language content into the context of use, and (ii) language production as an extraction of content from the context.[24] For these more demanding tasks, the CoL'89 approach, based on trees implemented as frames,[25] was not viable.

1.6 Comparison with Other Systems

To enable the DBS robot to freely communicate in natural language, we modeled it after the most evident properties of human language communication.[26] These have been reconstructed in the agent-oriented approach of the SLIM theory of language[27] (FoCL'99, NLC'06). SLIM provides the theoretical foundation for the innovations and adaptations of the DBS software machine.

The crucial issues distinguishing DBS from other current systems of natural language analysis may be summarized as follows:

---

[22] Thanks to Jaime Carbonell and the CMT/LTI at Carnegie Mellon University (1986–1989), who made programming the semantically interpreted CoL parser possible.

The 1983–1986 stay at Stanford University and the subsequent 1986–1988 stay at Carnegie Mellon University were supported by a five-year DFG Heisenberg grant. The 1988–1989 stay at CMU was supported by a Research Scientist position at the LCL (Dana Scott and David Evans). Thanks also to Brian MacWhinney, who supported a fruitful three-month stay at the CMU Psychology Department in the fall of 1989.

[23] Cf. FoCL'99, Sect. 21.4, for a review.

[24] See FoCL'99, Sect. 22.2, for a summary.

[25] In FrameKit, a desired value is retrieved by specifying the frame name, e.g., Fido, and the slot, e.g., has_brothers. The system returns the value(s) of the slot, e.g., Zach Eddie.

In DBS, a desired "next" proplet is activated by specifying its address in terms of the core and the prn value. These are provided by the "current" proplet. For example, navigating from the current proplet Fido to the successor proplet Zach would be based on the [has-brothers: Zach] feature and the prn value specified in the Fido proplet. The activation of the Zach proplet provides addresses for further successor proplets.

[26] A guiding principle is *form follows function* (Cuvier 1817/2009).

1.6.1 Summary of differences between DBS and other systems

1. **Derivation Order:**

   The parsing algorithm of DBS, i.e., LA-grammar, uses a strictly time-linear derivation order to compute *possible continuations*. The derivations of Phrase Structure Grammars and Categorial Grammars, in contrast, are partially ordered and compute *possible substitutions*. As a consequence, the application of LA-grammar to natural language has been shown to be of linear complexity, while the complexity of the other grammar formalisms is either polynomial but empirically inadequate (context-free), or computationally intractable ranging from context-sensitive (exponential) to recursively enumerable (undecidable; cf. FoCL'99, Chaps. 9–12).[28]

2. **Ontology:**

   In DBS, the model of natural language communication is located inside the speaker-hearer as a software machine attached to the agent's external and internal interfaces (cf. FoCL'99, Sect. 2.4). In Truth-Conditional Semantics, in contrast, the speaker-hearer is part of a set-theoretic model, and nonprocedural metalanguage definitions are used to connect the language expressions directly to referents in the model. As a consequence, DBS is designed for a talking robot, while Truth-Conditional Semantics is not[29] (cf. FoCL'99, Chaps. 19–21).

3. **Elementary Meaning:**

   In DBS, the agent's basic recognition and action procedures are reused as the elementary meanings of language. In Truth-Conditional Semantics, in contrast, elementary meanings are defined in terms of their truth-

---

[27] The acronym SLIM abbreviates Surface compositional, time-Linear, Internal Matching, which are the methodological, empirical, ontological, and functional principles of the approach. When the acronym is viewed as a word, SLIM indicates low mathematical complexity, resulting in the computational efficiency required for real-time processing.

[28] Tree Adjoining Grammar (TAG, Joshi 1969) parses "mildly" context-sensitive languages in polynomial time. For example, the formal languages $a^k b^k c^k$ (three terms), $a^k b^k c^k d^k$ (four terms), $a^k b^k c^k d^k e^k$ (five terms) are called mildly context-sensitive, but anything longer than five terms, e.g., $a^k b^k c^k d^k e^k f^k$, is called "strongly" context-sensitive and is of exponential complexity in TAG. The algorithm of LA-grammar, in contrast, parses all these formal languages in linear time, regardless of the number of terms.

[29] We are referring here to the classical approach of Truth-Conditional Semantics based on set theory, as in Montague grammar. This is in contradistinction to adaptations of Truth-Conditional Semantics to applications. For example, the robot STRIPS (Fikes and Nilsson 1971) uses the GPS (Global Positioning System) strategy for the motion planning of an artificial agent. The reasoning for this is formulated in terms of higher-order predicate calculus and "theorem proving," based on large numbers of "well-formed formulas" (wffs). To plan the motion of a robot like STRIPS, however, "certain modifications must be made to theorem-proving programs" (op. cit., p. 192).

   For a comparison of logical semantics and the semantics of programming languages, see FoCL'99, Sect. 19.2. Our alternative to the theorem proving method in AI is DBS inferencing. See 11.2.4 and 11.2.5 for a listing of The inferences defined in this book.

conditions relative to a set-theoretic model. As a consequence, the meanings in DBS have concrete realizations in terms of software and hardware procedures, while those of Truth-Conditional Semantics do not (cf. FoCL'99, Chaps. 3–6; NLC'06, Chap. 2).

4. **Database:**
In DBS, the content derived in the hear mode or by inferencing is stored in a *content-addressable* memory, called Word Bank. Most current applications, in contrast, use a *coordinate-addressable* database, for example, an RDBMS, if they use a database at all. The crucial property of content-addressable memories is that they are good for content which is written once and never changed. Given that a cognitive agent is constantly changing, this seems to be a paradoxical quality. It turns out, however, that it is the no-rewrite property which allows for a simple, powerful definition of inferences in DBS (Chaps. 5, 6, 10, 11).[30]

5. **Data Structure:**
DBS uses flat (non-recursive) feature structures with ordered attributes. Current systems of Nativism, in contrast, use recursive feature structures with unordered attributes to model "constituent structure" trees (cf. FoCL'99, Sect. 8.5). Flat feature structures with ordered attributes are of superior computational efficiency for a wide range of operations, such as pattern matching, which is ubiquitous in Database Semantics (Sects. 3.4, 4.2, 4.3, 5.2, 6.5).

6. **Intention**
DBS reconstructs the phenomenon of intention as part of an autonomous control designed to maintain the agent in a state of balance. This is in contrast to other schools of linguistics and philosophy who refer eclectically to Grice (1957, 1965, 1989) whenever the need arises, but are oblivious to the fact that Grice's elementary, atomic, presupposed notion is of little use for the computational reconstruction of intention and, by Grice's own definition (5.1.1), of meaning in a cognitive agent.

7. **Perspective**
The speak and the hear modes in the agent-oriented approach of DBS provide the foundation for modeling the perspectives of the speaker/writer and the hearer/reader in dialogue/text. They are (i) the perspective of an agent recording a current situation as a content, (ii) a speaker's perspective on a stored content, and (iii) the hearer's perspective on a content transmit-

---
[30] The additional means of coreference-by-address, coded both declaratively and as pointers, relates new content to old, satisfying the desiderata (i) of a declarative specification (using symbols) and (ii) of an efficient implementation (using pointers). Cf. Sect. 4.4.

ted by natural language (Chaps. 10, 11). In DBS, the computation of these perspectives is based on (i) suitable inferences and (ii) the values of the agent's STAR parameters, for Space, Time, Agent, Recipient.

The traditions from Structuralism to Nativism in linguistics and from Symbolic Logic to Truth-Conditional Semantics in analytic philosophy have not gotten around to addressing these points, let alone coming up with their own cogent model of natural language communication suitable for a talking robot.

It is therefore little wonder that engineers and researchers trying to satisfy an ever-growing demand for applications in human-machine communication are turning in droves to other venues. Having long given up on expecting efficient algorithms and sensible models[31] from linguistics and philosophy, the engineers see no other option than to rely on their technology alone, while many researchers have turned to *statistics* and *markup* by hand, as in statistical tagging and the manual metadata markup of corpora.[32]

Unfortunately, neither statistics nor markup seems to have a good long-term prospect for building a talking robot. What is needed instead is a computational reconstruction of cognition in terms of interfaces, components, functional flow, data structure, algorithm, database schema, and so on.

This in turn presupposes a theory of how natural language communication works. The crux is that communication includes language production, and language production requires an *autonomous control* for appropriate behavior. Where else should the contents come from which are realized by the speaker as language surfaces?

---

[31] Consider, for example, Kripke's (1972) celebrated theory of proper names, defined as "rigid designators" in a set-theoretic model structure, and try to convince someone to use it in a practical application.

[32] Markup and statistics are combined by using the manually annotated corpus as the *core corpus* for the statistical analysis, usually based on HMMs. Markup by hand requires the training of large personnel to consistently and sensibly annotate real texts. Statistical tagging can handle only a small percentage of the word form types, at low levels of linguistic detail and accuracy (cf. FoCL'99, Sect. 15.5), in part because about 50% of the word forms in a corpus occur no more than once (hapax legomena).

Corpora used for statistical analysis are big. A *de facto* standard is the British National Corpus (BNC, 1991–1994) with 100 million running word forms. Corpora marked up by hand are usually quite small. For example, MASC I (for Manually Annotated SubCorpus I) by Ide et al. (2010) consists of ca. 82 000 running word forms, which amounts to 0.08% of the BNC. Extensions of MASC I are intended as the core corpus for a statistical analysis of the American National Corpus (ANC).

An alternative approach are probabilistic context-free grammars (PCFGs) which use statistics to produce phrase structure and dependency trees. Examples are the Stanford parser (Klein et al. 2003, et seq.) and the Berkeley parser (Petrov et al. 2007, et seq.). An M.A. thesis at the CLUE currently investigates the possibility of using a PCFG parser as the front end for a broad coverage DBS system.

This experiment is based on reinterpreting the sign-oriented PCFG approach as the hear mode of an agent-oriented approach (Sect. 12.4). For use in DBS, the semantic relations coded indirectly (Sect. 7.1) by the context-free PCFG trees must be translated automatically into sets of proplets. The idea is a test of the storage, retrieval, inferencing, language interpretation, and language production capabilities of DBS with the large amounts of data provided by the PCFG parser.

To be effective, an agent's autonomous control should be based on flexible thought[33] rather than some rigid decision mechanism. In DBS, the analysis of thought proceeds from the following assumption:

Assumption: We can study thought by studying natural language.

This is not new, except that DBS starts from the *modality-dependent unanalyzed external surfaces* of natural language to model the transmission of information between the speaker and the hearer computationally.

The mechanism is based on pattern matching (Sect. 2.6) and works just as well for agent-external nonlanguage phenomena (L&I'05). Treating nonlanguage and language contents alike in recognition and action suggests treating them alike also in the think mode. This has the advantage that (i) they may be processed the same in a "horizontal" time-linear fashion and (ii) their "vertical" interaction for reference is facilitated (cf. FoCL'99, 5.4.1, 5.4.2).

The assumption of a close structural similarity between language and thought has the following consequences on the DBS software design:

Consequences:

1. Language content and nonlanguage content are *coded* practically the same, using flat feature structures (proplets) and addresses.
2. Language content and nonlanguage content are *stored* the same, using a content-addressable database (Word Bank).
3. Language content and nonlanguage content are *processed* the same, using a time-linear algorithm (LA-grammar).
4. The formal similarity between language content and nonlanguage (context) content is used for a cognitive reconstruction of *reference*.

Is it possible for such a design to be *psychologically real*? If humans and machines can communicate freely with nothing but time-linear sequences of modality-dependent unanalyzed external surfaces, then their cognitions may be equivalent – at certain levels of abstraction. Each equivalence level must be continuously verified by long-term upscaling for many different natural languages and applications, serving as a broad empirical foundation.

To give an idea of the functionalities demanded by a computational model of a cognitive agent with language, the following Part I is organized around five mysteries of natural language communication. Each mystery is answered with a Mechanism of Communication, MoC-1–MoC-5.

---

[33] The term *thought* is used here to refer to the cognitive processing of nonlanguage content, i.e., content independent of language surfaces.

Part I

**Five Mysteries of
Natural Language Communication**

2. Mystery Number One:
Using Unanalyzed External Surfaces

The first mystery of natural language communication may best be illustrated with a foreign language situation. For example, if our hometown is in an English-speaking country we can go to a restaurant there and successfully order a glass of water by saying to the waiter **Please bring me a glass of water**. If we travel to France, however, we will not be understood unless we use French or the waiter speaks English.[1]

2.1 Structure of Words

This difference between speaking in our language at home and abroad is caused by the composite structure of words. Essential components are (i) the *surface*, (ii) the *meaning*, and (iii) a *convention* connecting the surface and the meaning. As informal examples, consider the following analyses of the English word **water** and its French counterpart **eau**:

2.1.1 INFORMAL EXAMPLES SHOWING BASIC WORD STRUCTURE

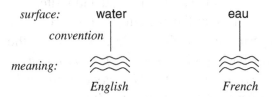

The two words have the same literal meaning, represented by three wavy lines suggesting water. Their surfaces, however, are the different letter sequences **water** and **eau**. Each surface is connected to its meaning by a convention which every speaker of English or French has to learn.[2]

---

[1] "Whereas the individuals of all nonhuman species can communicate effectively with all their conspecifics, human beings can communicate effectively only with other persons who have grown up in their same linguistic community – typically, in the same geographical region." Tomasello (2003), p. 1.

[2] The Swiss linguist Ferdinand de Saussure (1857–1913) describes the convention-based and therefore "unmotivated" relation between language-dependent surfaces like **water** or **eau** and their meaning in his *Premier Principe: l'arbitraire du signe* (Saussure 1916/1972; FoCL'99, Sect. 6.2).

We assume that a word form has an abstract surface type. This type may be realized in different modalities as tokens, mainly in spoken or written language.[3] In the modality of spoken language, the surfaces are sounds which are recognized by the agent's ears and produced by the agent's mouth. In the modality of written language, the surfaces are letter sequences which are recognized by the agent's eyes and produced by the agent's hands.

It follows from the basic structure of words that first language acquisition requires the child to learn (i) the meanings, (ii) the (acoustic) surfaces, and (iii) the conventions which connect the surfaces to the correct meanings.[4] This process is embedded in child development, takes several years, and normally does not cause any special difficulties. However, when learning the words of a foreign language as an adult, the following difficulties stand out:

2.1.2 TASKS OF LEARNING THE WORDS OF A FOREIGN LANGUAGE

- learning to recognize and produce the foreign surfaces in the modalities of spoken and written language, and

- learning the conventional connections between the foreign surfaces and meanings familiar from one's first language.

Learning to recognize the acoustic surfaces of a foreign language is difficult and to pronounce them without accent is often nearly impossible, while learning to read and to write may come easier. There are languages like Japanese, however, for which a Westerner is considered more likely to learn to speak fairly fluently than to acquire a near-native ability to read and write.

Connecting the foreign surfaces to familiar meanings presupposes that the notions of the foreign language are identical or at least similar to those of one's own. This holds easily for basic notions such as *father, mother,*[5] *child, son,* and *daughter*, as well as *sun, moon, water, fire, stone, meat, fish, bird,* and *tree*. When it comes to more culturally dependent notions, however, what is represented by a single word (root) in one language may have to be paraphrased by complex constructions in the other.

---

[3] A third modality is signed language for the hearing impaired and, as a form of written language, there is Braille for the blind. For the type-token distinction see NLC'06, Sect. 4.2; FoCL'99, Sect. 3.3.

[4] In addition, the child has to learn the syntactic-semantic composition of words into complex expressions (sentences) with complex meanings and their use conditions.

[5] Kemmer (2003, p. 93), claims that "in some languages" the word meaning of **mother** would include maternal aunts. It seems doubtful, however, that the distinction between a mother's own children and those of her sisters could ever be lost. A more plausible explanation is a nonliteral use. In German, for example, a child may call any female friend of the parents **Tante** (aunt). If needed, this use may be specified more precisely as **Nenn-Tante** (aunt by name or by courtesy). Similarly in Korean, where the term "older brother" may be bestowed on any respected older male (Kiyong Lee, personal communication).

A popular example is the Eskimo language Inuit, said to have something like fifty different words for snow (Boas 1911). In English translation, these would have to be paraphrased, for example, as **soft snow, hard snow, fresh snow, old snow, white snow, grey snow,** and **snow pissed on by a baby seal.**[6] A precondition for adequate paraphrasing is a proper knowledge of the foreign notion to be described. As an example consider the first Eskimo faced with the task of introducing the notion of an electric coffee grinder into Inuit.

2.2 Modality-Dependent Unanalyzed External Surfaces

As an agent-oriented approach, DBS provides two basic perspectives from which communicating with natural language may be analyzed, namely *internal* and *external*. The internal (or endosystem) perspective is based on one's introspection as a native speaker, and has been illustrated in the previous section by an example of communication failure in a foreign language environment. The external (or exosystem) perspective is that of a scientist (cf. NLC'06, Sect 1.4) working to reconstruct the language communication mechanism as an abstract theory, verified by an implementation as an artificial agent with language (talking robot).

The following example illustrates the external perspective with two agents, A and B. A is in the speak mode and produces the modality-dependent unanalyzed external word form surface **water**. B is in the hear mode and recognizes this surface.[7]

2.2.1 PRODUCTION AND RECOGNITION OF A WORD

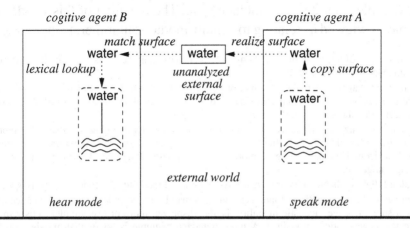

[6] See Pullum (1991), "The great Eskimo vocabulary hoax."
[7] For a refined version see 2.6.1.

The two agents, represented as boxes, are concrete individuals with bodies[8] existing in the real world. The external surface of the word **water** is shown as a letter sequence between the two agents (modality of vision).

Such an external surface has neither a meaning attached to it nor any grammatical properties. It is simply an external object with a particular shape in a certain modality, which may be measured and described with the methods of the natural sciences, e.g., formant structure. The shape is arbitrary in the sense that it does not matter whether the agents use the surface **water** or the surface **eau** as long as they obey the conventions of their common natural language.

In their minds, in contrast, both agents contain the word **water** as analyzed in 2.1.1, i.e., the surface, the meaning, and the convention-based connection between the surface and the meaning. As purely cognitive representations (e.g., binary code), the analyzed words are independent of any modality,[9] but include the meanings and all the grammatical properties.

The speaker realizes the external surface by copying the internal word surface and performing it externally in the modality of choice (*copy-realize*). The hearer recognizes the external surface by matching it with a learned surface and by using the recognized surface for lexical lookup (*match-lookup*).[10]

Even though the internal representations of surfaces and meanings are modality-free[11] in principle, they may relate indirectly to the modality of the external interface by which they have been recognized or realized. During recognition an internal surface matches a modality-dependent representation such as sound, and during production it is realized in such a modality.

In this indirect sense, the internal surfaces are usually mono-modal[12] while the meanings are multi-modal. For example, the surface of the word **raspberry** is mono-modal in that the associated modality is purely acoustic, optical, or tactile (Braille), but does not include taste. The meaning of this word, in contrast, may be viewed as relating to a multi-modal conglomerate including taste.

---

[8] The importance of agents with a real body (instead of virtual agents) has been emphasized by Emergentism (MacWhinney 1999, 2008).

[9] To illustrate the agent-internal representation of the word **water** in 2.2.1, we had to resort to a suitable modality, namely vision. Though unavoidable, this is paradoxical insofar as the words in the cognition of an agent are inherently modality-free.

[10] One may recognize a surface as a real word, yet not know its meaning, for example "reprobate," "exigency," "prevaricate," or "effloresce" in English. In this case, there is an entry missing in the hearer's lexicon. This may hold even more when recognizing a surface-like shape as a word form of a foreign language.

[11] Barsalou (1999) calls modality-free *modality-general* and modality-dependent *modality-specific*.

[12] A multi-modal representation of language surfaces may be found in an opera performance in which the text sung on stage (acoustic modality) is also displayed more or less simultaneously in writing above the stage (optical modality). A more mundane example is an English movie with English subtitles.

The fundamental basis of the information transfer mechanism of natural language is formulated as the First Mechanism of Communication:

2.2.2 THE FIRST MECHANISM OF COMMUNICATION (MOC-1)

Natural language communication relies on modality-dependent *external surfaces* which are linguistically unanalyzed in that they have neither meaning nor any grammatical property.

MoC-1 is essential for the construction of DBS robots, and differs from the assumptions of "Realism," e.g., Situation Semantics (Barwise and Perry 1983, cf. p. 226), and of Truth-Conditional Semantics (e.g., Montague 1974).

Given that the external surfaces are modality-dependent, while their internal counterparts are modality-free, there is a constant mapping between modality-dependent and modality-free representations during communication (Conversion Universal 2 in 1.1.3). More specifically, in the speak mode, the mapping is from modality-free analyzed internal surfaces to modality-dependent, unanalyzed external surface representations. In the hear mode, the mapping is from modality-dependent unanalyzed external surfaces to modality-free analyzed internal surface representations.

MoC-1 is a functional complement to Surface Compositionality.[13] According to this methodological principle, the grammatical analysis of language signs may use only the word forms of concrete surfaces as the building blocks of composition, such that all syntactic and semantic properties of a complex expression derive systematically from the syntactic category and the literal meaning of the lexical items (which are of a cognitive, modality-free nature).

It follows from MoC-1 that reconstructing the mechanism of natural language communication cannot be limited to a grammatical analysis of isolated language signs, but must include a functional model, defined as follows:

2.2.3 FUNCTIONAL MODEL OF NATURAL LANGUAGE COMMUNICATION

A functional model of natural language communication requires

1. a set of cognitive agents each with (i) a body, (ii) external interfaces for recognition and action, and (iii) a memory for the storage and processing of content,
2. a set of external language surfaces which can be recognized and produced by these agents by means of their external interfaces using pattern matching,

---

[13] Cf. SCG'84; FoCL'99, pp. 80, 111, 256, 327, 418, 501, 502; NLC'06, pp. 17–19, 29, 89, 211.

3. a set of agent-internal (cognitive) surface-meaning pairs established by convention and stored in memory, whereby the internal surfaces correspond to the external ones, and

4. an agent-internal algorithm which constructs complex meanings from elementary ones by establishing semantic relations between them.

The definition implies that the cognition of agents in a language community is as important for the definition of natural language as the unanalyzed external surfaces. This may be shown by "lost languages" (FoCL'99, Sect. 5.3).

Imagine the discovery of clay tablets left behind by an unknown people perished long ago. Even though the external surfaces of their language are still present in form of the glyphs on the tablets, the language as a means of communicating meaningful content is lost. The only way to revive the language at least in part is to reconstruct the *knowledge* of the original speakers – which was part of their cognition and is now part of the cognition of the scientists.[14]

The requirements of 2.2.3 are minimal. They are sufficient, however, to distinguish natural language communication from other forms:

2.2.4 FORMS OF COMMUNICATION WITHOUT NATURAL LANGUAGE

- endocrinic messaging by means of hormones,
- exocrinic messaging by means of pheromones, for example in ants,[15] and
- the use of samples, for example in bees communicating a source of pollen.

These forms of communication differ from natural language because they lack a set of external surfaces with corresponding internal surface-meaning pairs established by convention.[16]

From a functional point of view, the mechanism described by MoC-1 has the following advantages for communication:

2.2.5 ADVANTAGES FOLLOWING FROM MoC-1

1. The modality-free internal meanings attached to the internal surfaces are not restricted by the modality of the external surfaces.

---

[14] Thereby, knowledge of the word form meanings and the algorithm for their composition alone is not sufficient for a complete, successful interpretation. What is needed in addition is the archaeologists' knowledge of the context of use.

[15] E. O. Wilson (1998, p. 229) describes the body of an ant worker as "a walking battery of exocrinic glands."

[16] Another example of not constituting a language is the macros used in programming languages. Defined ad hoc by the programmer as names for pieces of code, they are (program-)internal abbreviations. As such they do not require any external surfaces agreed on by convention and are not used for inter-agent communication. Nevertheless, as abbreviation devices, macros model an important ability of natural language.

2. The modality-dependent external surfaces are much better suited for (i) content transfer and (ii) agent-external long-term storage than the associated agent-internal modality-free meanings.

The advantage of modality-free meanings referred to in (1) may be illustrated by a sign-theoretic comparison of symbols and icons (cf. FoCL'99, Sect. 6.4). Icons in the visual modality, for example, are limited to a visual representation of meaning, which makes it often impossible to represent concepts from another modality, for example, the meaning of *sweet*. Symbols, in contrast, have no such limitation because their meaning representations are modality-free.[17]

The advantage of modality-dependent external surfaces for transfer and storage referred to in (2) is that an unanalyzed external surface is simple and robust compared to the often complicated meanings, attached by convention to the surfaces inside the cognitive agent. This holds especially for the written representation of language, which has long been the medium of choice for the agent-external storage of content.[18]

Naturally, the highly specialized and powerful technique characterized by MoC-1 also has an apparent disadvantage: because the surface for a given meaning can take any shape within the limits of its modality, communities of natural agents can and do evolve their own languages. This results in the difficulty of communicating in foreign language environments, which this chapter began with.

2.3 Modality Conversion in the Speak and Hear Modes

MoC-1 is not only a conceptual insight into the working of natural language communication between agents viewed from the outside, but constitutes a well-defined technical challenge, namely the construction of machines which can recognize and produce external language surfaces. The basic task of these machines may be viewed as an automatic conversion between a modality-dependent realization of an external surface and its modality-free counterpart

---

[17] A modality-free representation of meaning is procedural in the sense that it is based on the recognition and action procedures of the cognitive agent. Whether such a meaning representation in an artificial agent is adequate or not is decided by the agent's performance. For example, an artificial agent's concept of *shoe* may be considered adequate if the agent picks out the same object(s) from a collection of different things as a human would (NLC'06, Chap. 4).

[18] Recently, storage by means of written language has been complemented by the technologies of tape recording, video, CDs, DVDs, etc., which allow to preserve also spoken and signed language for indefinite periods of time.

On the relation between modalities and media see FoCL'99, pp. 23 f., and NLC'06, pp. 23 f.

represented as agent-internal digital code, e.g., seven-bit ASCII. This conversion is instantiated in four basic variants, namely the speak mode and the hear mode each in the modalities of vision (optics, writing) and audition (acoustics, speech).

In the hear mode, today's systems of *speech recognition* convert external acoustic surfaces into modality-free digital code, mainly by the statistical method of Hidden Markov Models (HMMs). In the optical modality, today's systems turn images of letters into modality-free digital code based on the software of *optical character recognition* (OCR).

In the speak mode, today's systems of *speech synthesis* convert digitally represented text into artificially realized speech, usually by concatenating pieces of recorded speech. In the modality of vision, there is the conversion from digital code to the familiar letter images on our computer screens, which may be called *optical character synthesis* (OCS).

The conversion from a modality-dependent to a modality-free representation (hear mode) abstracts away from properties of the external surfaces such as speed, pitch, intonation, etc. in spoken language, and font, size, color, etc. in written language, which from a certain point of view may be regarded as accidental. The result of this token-type mapping is a *surface template* (2.3.1, 2.6.1). Conversely, the conversion from a modality-free to a modality-dependent representation (speak mode) must settle on which of these properties, for example, pitch, speed, dialect, etc., should be selected for the external surface (type-token mapping).

If speak mode and hear mode utilize the *same* modality and are realized by *different* agents, we have inter-agent communication. Examples are agent A writing a letter (speak mode, visual modality) and agent B reading the letter (hear mode, visual modality), and accordingly in the auditory modality:

2.3.1 INTER-AGENT COMMUNICATION USING SPEECH

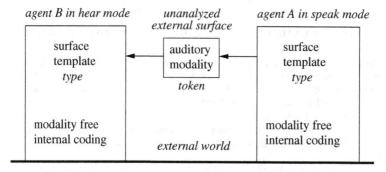

A notable difference between the auditory and the visual modality in inter-agent communication is that the interpretation(s) (read, hear mode) of a letter, for example, may be far removed in time and space from the point of production (write, speak mode).[19] Another difference is that written language can be corrected by the author, while spontaneous speech cannot:

> Speech is irreversible. That is its fatality. What has been said cannot be unsaid, except by adding to it: to correct here is, oddly enough, to continue.
>
> R. Barthes (1986, p. 76)

Of course, as soon as spoken language is recorded, an interpretation may be arbitrarily distant in time and space from the production. Also, a recording may be "doctored" – which from a certain point of view is a form of correction.

If the speak mode and the hear mode utilize *different* modalities and are realized by the *same* agent, we have a modality conversion. In nature, modality conversion is illustrated by reading aloud (conversion from the visual to the auditory modality) and taking dictation (conversion from the auditory to the visual modality).

2.3.2 TWO KINDS OF MODALITY CONVERSION

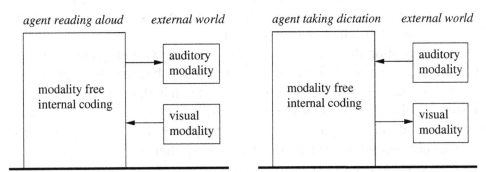

In technology, reading aloud is modeled by combining optical character recognition and speech synthesis, which is an important application for the blind. Conversely, taking dictation is modeled by a machine called "electronic secretary," which combines speech recognition with optical character synthesis.[20]

---

[19] As shown in Chap. 11, the content of a text must be interpreted relative to the STAR-1 (Space, Time, Agent, Recipient) values of the author when producing the written sign. Because the reader is normally not present when a text is written – in contradistinction to the hearer in face-to-face communication – it is in the interest of the writer to code the STAR-1 into the text itself. Otherwise, the reader will have difficulty anchoring the written sign to its context of use.

[20] Infra-red cameras are another technical means of modality conversion, representing temperature (what is called *temperature modality*, Dodt and Zotterman 1952) as color (visual modality).

2.4 General-Purpose Interfaces and the Language Channels

In recognition, language interpretation is embedded into nonlanguage recognition. For example, in the visual modality humans can recognize nonlanguage[21] input such as streets, houses, cars, trees, and other agents as well as language input such as shop signs and text in newspapers, books, and letters. Similarly in the auditory modality: humans can recognize nonlanguage input such as the sounds made by rain, wind, birds, or approaching cars, as well as language input such as speech from other agents, radio, and television.

In action, language production is likewise embedded into nonlanguage action. For example, humans can use hands for generating nonlanguage output, such as filling the dishwasher or for drawing a picture, as well as for language production in the visual modality, such as writing a letter or typing something on a computer keyboard. Similarly in the auditory modality: humans can use their voice for nonlanguage output, such as for singing without words or for making noises, as well as for language production, i.e., for speech.

Such embedding of language recognition and synthesis into nonlanguage recognition and action cannot be handled by today's technologies of speech recognition and optical character recognition. Instead, they use input and output channels dedicated to language. This constitutes a substantial simplification of their respective tasks (smart solution).[22]

Despite this simplification and well-funded research with many promises and predictions over several decades, automatic speech recognition has not yet reached its goal.[23] For proof, we do not have to embark upon any argument, but simply point to the ever-increasing number of keyboards in everyday use: if automatic speech recognition worked in any practical way, i.e., comparably to the speech recognition capability of an average human,[24] few users would prefer the keyboard and screen over speech.

For building a talking robot like C3PO any time soon, insufficient automatic speech recognition presents an – almost certainly temporary – obstacle.

---

[21] We prefer the term nonlanguage over nonverbal because the latter leads to confusion regarding the parts of speech, e.g., nouns, verbs, and adjectives.

[22] For the distinction between smart and solid solutions, see FoCL'99, Sect. 2.3

[23] Optical character recognition, in contrast, is quite successful, especially for printed language, and widely used for making paper documents available online.

[24] Technically speaking, a practical system of speech recognition must fulfill the following requirements simultaneously (cf. Zue et al. 1995):
   1. speaker independence,
   2. continuous speech,
   3. domain independence,
   4. realistic vocabulary,
   5. robustness.
No system of speech recognition has yet achieved these desiderata simultaneously.

But it does not hinder the current work on the computational reconstruction of natural language communication in DBS. The reason is an important difference between natural and artificial cognitive agents regarding what we call the *auto-channel* and the *service channel*.[25]

A natural agent has only an auto-channel for recognition and action. The auto-channel evolves naturally during child development and comprises everything a natural agent can see, hear, feel, taste, etc., as well as consciously do. The auto-channel also includes natural speech recognition and production.

In a standard computer, e.g., a notebook or a desktop computer, in contrast, there is no auto-channel. Instead, there is only a service channel consisting of the keyboard and the screen. It allows users and scientists alike to access the hardware and software of the computer directly.

Building an artificial cognitive agent consists largely of reconstructing an artificial auto-channel, with external interfaces for vision, audio, locomotion, manipulation, etc. In higher cognitive agents, the auto-channel is connected to the agent's autonomous control (Chap. 5).

The computational reconstruction of the auto-channel is an incremental process which relies heavily on the keyboard and the screen. They function as the service channel for direct access to, and manipulation of, the hardware and software of the robot under development. After completion, the artificial agent will be able to interact autonomously with the external world via its auto-channel, including communication with the human user. However, in contrast to a natural agent, an artificial agent will not only have an auto-channel, but also a service channel as a remnant of the process of its construction.

The essential role of the service channel in bootstrapping the reconstruction of cognition is especially clear in the area of natural language communication. This is because it does not really matter whether a language surface gets into the computer via the auto-channel or via the service channel. All that matters for modeling natural language understanding in computational linguistics is that the word form surfaces get into the computer, and for this typing them at the keyboard of today's standard computers is sufficient.

Similarly for language production in the speak mode: all that matters for the user's communication with the computer is that the surfaces derived from prior cognitive processing be realized externally, and for this optical character synthesis on the screen of today's standard computers is sufficient – at least for users who can see and have learned to read.

The possibility to reconstruct natural language processing in artificial agents via the service channel is good news for developing capable automatic speech

---

[25] Cf. NLC'06, pp. 14 f.

recognition[26] as part of the artificial auto-channel. The reason for a 20 year stagnation in this area is a search space too large for a statistical approach as used today. The gigantic size of the search space is caused by the large number of possible word forms in a natural language multiplied with an even larger number of possible syntactic combinations and variations of pronunciation between different speakers.

The best way to reduce this search space is by hypotheses on possible continuations computed by a time-linear grammar algorithm as well as by expectations based on experience and general world knowledge. After all, this is also the method used by humans for disambiguating speech in noisy environments.[27] Using this method for artificial speech recognition requires a theory of how communicating with natural language works.

2.5 Automatic Word Form Recognition

Assuming that the language input to an artificial cognitive agent is word form surfaces provided by the service channel as sequences of letters, the first step of any rule-based (i.e., nonstatistical) reconstruction of natural language understanding[28] is building a system of automatic word form recognition. This is necessary because for the computer a word form surface like **learns** is merely a sequence of letters coded in seven-bit ASCII, no different from the inverse letter sequence **snrael**, for example.

Automatic word form recognition takes an unanalyzed surface, e.g., a letter sequence like **learns**, as input and provides the computer with the information needed for syntactic-semantic processing. For this, any system of automatic word form recognition must provide (i) *categorization* and (ii) *lemmatization*. Categorization specifies the grammatical properties, which in the case of **learns** would be something like "verb, third person singular, present tense." Lemmatization specifies the base form, here **learn**, which is used to look up the meaning common to all the word forms of the paradigm, i.e., **learn, learns, learned,** and **learning.**[29]

---

[26] Contextual cognition, such as nonlanguage vision and audio, may also benefit from the service channel. By building a context component with a data structure, an algorithm, and a database schema via direct access, the robot's recognition and action components are provided with the structures to map into and out of.

[27] Juravsky and Martin (2009, pp. 16–17) sketch an approach which could be construed as being similar to DBS. However, their system is not agent-oriented, nor surface compositional, nor time-linear.

[28] We begin with the hear mode as a means to get content into the computational reconstruction of central cognition. The availability of such content is a precondition for implementing the speak mode.

[29] For further information on the morphological analysis of word forms and different methods of automatic word form recognition, see FoCL'99, Chaps. 13–15.

The recognition algorithm in its most primitive form consists of matching the surface of the unknown letter sequence with the corresponding surface (key) in a full-form lexicon, thus providing access to the relevant lexical description.

2.5.1 MATCHING AN UNANALYZED SURFACE ONTO A KEY

      unanalyzed word form surface: learns

                                            matching

      morphosyntactic analysis: [learn/s, categorization, lemmatization]

In DBS, the recognition algorithm consists of (i) the segmentation of the letter sequence of a surface into known but unanalyzed parts, (ii) lexical lookup of these parts in a trie structure,[30] and (iii) composition of the analyzed parts into well-formed analyzed word forms (cf. FoCL'99, Chap. 14). This requires (i) an online lexicon for base forms, (ii) allo-rules for deriving different variants of a morpheme, e.g., wolf and wolv-, before runtime, and (iii) combi-rules for combining the analyzed allomorphs during runtime.

Building such a system of automatic word form recognition for any given natural language is not particularly difficult, even for writing systems based on characters, e.g., Chinese and Japanese, rather than letters. Given (i) an online dictionary of the natural language of choice, (ii) a suitable off-the-shelf software framework, and (iii) a properly trained computational linguist, an initial system can be completed in less than six months.[31] It will provide accurate, highly detailed analyses of about 90% of the word form types in a corpus.

Increasing the recognition rate to approximately 100% is merely a matter of additional work.[32] It consists of adding missing entries to the online lexicon, and improving the rules for allomorphy and for inflection or agglutination, derivation, and composition. To maintain a recognition rate of practically 100% over longer periods of time, the system must be serviced continually, based on a RMD corpus, i.e. a Reference Monitor corpus with a Domain structure (Sect. 12.2).

2.6 Backbone of the Communication Cycle

Automatic word form recognition is the first step of natural language interpretation in the hear mode. Automatic word form synthesis is the last step of

---

[30] See Knuth (1998, pp. 495–512).

[31] This is the standard period of time for writing an MA thesis at the University of Erlangen-Nürnberg.

[32] This is in contrast to the statistical method, which does not lend itself to the correction of specific errors. See FoCL'99, Sect. 15.5.

language production in the speak mode. The two mechanisms are alike in that they establish a correlation between unanalyzed surfaces and analyzed surfaces (word forms) by means of pattern matching, but differ in the direction:

2.6.1 BACKBONE OF SURFACE-BASED INFORMATION TRANSFER

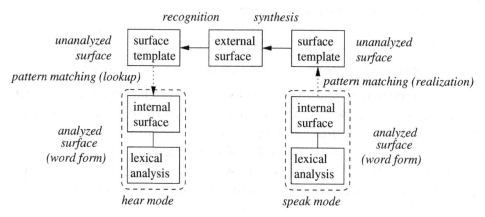

At the center of an inter-agent information transfer by means of natural language is the unanalyzed *external surface*. It is the output of the speak mode and the input to the hear mode. The existence of an external surface may be shown objectively by the natural sciences measuring it in its modality.

In the speak mode, the internal surface is mapped into a template used to realize an external surface (synthesis). In the hear mode, the external surface is recognized by matching it with an internal template used to retrieve a corresponding analyzed word form (lexical lookup). The surface templates are modality-dependent, while the internal surfaces are modality-free.

In successful communication, the analyzed word form must be the same in the speaker A and the hearer B. This requires that they associate the internal surface with the same lexical analysis, i.e., A and B must have learned the same language. In addition, the word forms are assumed to match an agent-internal context of interpretation (not shown; cf. 4.3.2 and 4.3.3).

After lexical lookup in the hear mode and before word form synthesis in the speak mode, the cognitive processing of content in the think mode is completely independent of the external interfaces of the agent and the external surfaces. In the speak and hear modes, however, central cognition must work hand in glove with the external interfaces of the agent.

For unrestricted human-machine communication, the interaction between interfaces and cognition should be the same in the human prototype and the artificial agent. This has been preformulated in NLC'06, Sect. 1.5, as the following principle:

2.6.2 INTERFACE EQUIVALENCE PRINCIPLE

> For unrestricted human-machine communication, the artificial cognitive agent with language (DBS robot) must be equipped with the same interfaces to the external environment as the human prototype.

This principle follows from the modality-dependency of the external surfaces and the associated interfaces. The external interfaces of humans are concretely given and are therefore susceptible to an objective structural and functional analysis. Using a combination of the natural sciences and engineering, such an analysis helps to reconstruct the external interfaces of the artificial agent.

Another equivalence principle, also preformulated in NLC'06, Sect. 1.5, applies to the processing of the surfaces by the artificial agent:

2.6.3 INPUT/OUTPUT EQUIVALENCE PRINCIPLE

> In natural language interpretation, the artificial agent must analyze the modality-dependent unanalyzed external surface by
>
> (i)   segmenting it in the same way into parts (i.e., word forms) and
>
> (ii) ordering the parts in the same way (i.e., in a time-linear sequence)
>
> as the human prototype; and accordingly for language production.

Failure to establish segmentation and order correctly may disrupt communication. Designing a talking robot without integrating 2.6.3 would violate a basic rule of science and compromise the agent's functionality from the outset.

The Input/Output Equivalence Principle extends the backbone of surface-based information transfer, 2.6.1, from single word forms to complex expressions: for producing the external surface of a complex expression, the speaker's cognition must provide a sequence of analyzed internal word forms; for interpreting a sequence of unanalyzed external word form surfaces, the hearer's cognition must reconstruct a corresponding sequence of analyzed internal word forms, at least to the extent of lexical lookup.

An external time-linear sequence as the output of language production and as the input of language interpretation suggests a time-linear derivation order for the speak and the hear modes. For the cognitive connection between the two, i.e., the think mode, DBS adopts the time-linear derivation order as well.[33]

The time-linear design, motivated by simplicity and efficiency, raises the question: does it work empirically? So far, the DBS analysis of different gram-

---

[33] In this way, a constant switching between (i) the time-linear computing of possible continuations in the speak and hear modes and (ii) a computing of possible substitutions in the think mode is avoided.

matical constructions, such as relative clauses, gapping, bare infinitives, long-distance dependencies, prepositional objects, copula constructions, etc. (cf. Part II) in many different languages, has not discovered any empirical difficulty (fatal flaw) resulting from the use of a strictly time-linear derivation order for the full cycle of natural language communication.

This is in contrast to Phrase Structure Grammar and Truth-Conditional Semantics. Each has a fatal flaw when applied to natural language. Context-free Phrase Structure Grammar cannot handle the syntactic phenomenon of discontinuous elements, shown by Bar-Hillel (1953) (cf. FoCL'99, Chap. 8, Constituent Structure paradox). Truth-Conditional Semantics is inherently incapable of a complete analysis of natural language, shown by Tarski (1935) based on the Epimenides paradox (cf. FoCL'99, Sect. 19.5).

In DBS, the Constituent Structure paradox does not arise because the linguistic analysis is based on valency canceling in a time-linear derivation order (cf. FoCL'99, 10.5.4), and not on constituent structure. The Epimenides paradox is avoided in DBS because meanings are based on the agent's recognition and action procedures, and not on truth conditions.

Despite these long-standing results, Truth-Conditional Semantics and context-free Phrase Structure Grammars continue to be the formal systems of choice for today's mainstream approaches to natural language analysis. Simply ignoring their flaws does not make them any less fatal, however. Instead, they go a long way towards justifying the innovations and adaptations of DBS.

As the first, and so far the only, reconstruction of the cycle of natural language communication, the agent-oriented approach of DBS is based on the following innovations and adaptations:

1. the time-linear algorithm of LA-grammar,

2. the content-addressable memory of a Word Bank,

3. the data structure of proplets as non-recursive feature structures with ordered attributes,

4. the order-free coding of semantic relations between proplets by means of addresses,

5. the reconstruction of reference as an agent-internal correlation between language and context,

6. the reconstruction of intention in terms of maintaining the agent in a state of balance, and

7. the use of "/," "\," "|," and "−" edges to characterize functor-argument and coordination graphically.

These have created a divide between DBS on the one hand and the coalition of sign-oriented schools of Nativism and their allies from Truth-Conditional Semantics on the other. Yet the surface-based information transfer mechanism, 2.6.1, in combination with the principles 2.6.2 and 2.6.3 constitutes a simple, effective, and natural method no freely communicating robot can do without. Once accepted, the transfer mechanism sets into motion a chain of consequences which are presented here as the software model of the DBS robot.

3. Mystery Number Two:
Natural Language Communication Cycle

The second mystery is the compositional semantics of natural language. How are lexically analyzed word forms connected into content in the hear mode? How is content activated selectively in the think mode? How is activated content realized as external surfaces in the speak mode? To answer these questions, let us begin by further refining the format of analyzed word forms.

3.1 Choosing the Data Structure

From a software engineering point of view, choosing the format and the interpretation of the basic units for modeling the cycle of natural language communication amounts to choosing a certain data structure.[1] A well-defined data structure is required for specifying the input and output interfaces of the algorithm intended to operate on the data items of the system, as in their storage, processing, and memory retrieval.

The items recognized and produced by the cognitive agent's language interfaces are the external surfaces. Therefore, internal word forms are the natural candidates for being defined as the basic data structure. The format, called *proplet*, develops from more intuitive representations:

3.1.1 DEVELOPMENT OF THE PROPLET FORMAT

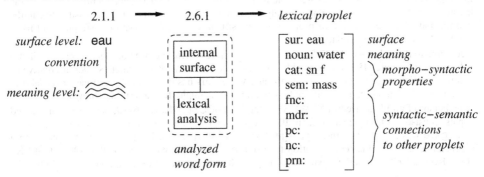

---
[1] In early computer science, the term "data structure" referred to hardware constructs such as queue, stack, linked list, car, cdr, array, or pointer. The development of higher programming languages

Representation 2.1.1, inspired by de Saussure, is made more specific in 2.6.1
by stipulating that the surface is agent-internal, and made more general by using **lexical analysis** to represent the meaning plus the morphosyntactic properties. The formats 2.1.1 and 2.6.1 are alike in that they use graphical means
(vertical line) to indicate the conventional connection between the surface and
the lexical properties.

The *lexical proplet* in 3.1.1 differentiates the analysis further by using the
format of non-recursive feature structures with ordered[2] attribute-value pairs
(**avp**s). This simple format is (i) suitable for storage and retrieval in a content-
addressable database, and (ii) allows us to code any interproplet relations, any
morphosyntactic properties, and the convention-based (unmotivated) connection between the internal surface and the lexical analysis.[3]

In a proplet, the lexical and the compositional aspects of a word form are
systematically distinguished by means of characteristic attributes. For example, the lexical aspect in the proplet shown in 3.1.1 is represented by the values
of the first four attributes: 1. [sur: eau] specifies the *surface*; 2. [noun: water]
consists of the *core attribute* specifying the part of speech, and the *core value*
specifying the literal meaning represented by a concept name; 3. [cat: sn f]
(for *category: singular noun, feminine*) specifies the morphosyntactic properties contributing directly to the combinatorics, for example, regarding agreement; 4. [sem: mass] (for *semantics: mass*[4]) specifies morphosyntactic properties not directly contributing to the combinatorics.

The compositional aspect of the word form is represented by the remaining attributes. Because the proplet illustrated in 3.1.1 is a lexical proplet, the
compositional attributes have no values yet, but will receive their values dur-

---

brought into use the term "abstract data type" to refer to "software" data structures, which are defined independently of their hardware implementation. With the increasing abstracting away from the
hardware level, the two terms are now widely used interchangeably. This makes it possible to call the
non-recursive feature structure of proplets a "data structure," which is a shorter, simpler, and more
intuitive term than "abstract data type."

[2] According to ISO 24610-1 (TC37/SC4), a standard feature structure is a(n unordered) *set* of attribute-
value pairs (Carpenter 1992), presumably for reasons of mathematical aesthetics. Our alternative of
defining proplets as a(n ordered) *list* of attribute-value pairs has the advantages of easier readability
and a more efficient computational implementation, for example, for pattern matching, cf. 1.3.1.

[3] A precursor of proplets, used in NEWCAT'86 and in FoCL'99, Chaps. 16–18, is a format of ordered
triples, consisting of (i) the surface, (ii) the category, and (iii) the base form. For example, the word
form **gave** was analyzed as the triple (**gave (n' d' a' v) give**), which may be transformed into a
character-separated value (csv) notation (as used in the original WordNet (Miller1996), CELEX2
(Olac Record 1996), and other lexical encodings). The corresponding proplet notation would be

$$\begin{bmatrix} \text{sur: gave} \\ \text{verb: give} \\ \text{cat: n' d' a' v} \\ \text{...} \end{bmatrix}$$, which is a flat (non-recursive) feature structure.

[4] In contradistinction to a discrete object, represented by the value **count**.

ing syntactic-semantic parsing in the hear mode (3.3.1). The *continuation* attributes fnc (for functor) and mdr (for modifier) will characterize the relations of the proplet to other proplets in terms of functor-argument, while pc and nc (for previous and next conjunct) will characterize relations of coordination. The value of the *book-keeping* attribute prn (for proposition number), finally, will be a number common to all proplets belonging to the same elementary proposition and also indicate the temporal order relative to other propositions.

3.2 Representing Content

One requirement on proplets as a data structure is that they must provide for a simple representation of propositional content. This applies specifically to the compositional semantics of *functor-argument* and *coordination*, intra- and extrapropositionally. Consider the content of an intrapropositional functor-argument, represented as a set of linked proplets:

3.2.1 FUNCTOR-ARGUMENT OF Julia knows John.

$$
\begin{bmatrix} \text{noun: Julia} \\ \text{fnc: know} \\ \text{prn: 625} \end{bmatrix}
\begin{bmatrix} \text{verb: know} \\ \text{arg: Julia John} \\ \text{prn: 625} \end{bmatrix}
\begin{bmatrix} \text{noun: John} \\ \text{fnc: know} \\ \text{prn: 625} \end{bmatrix}
$$

The simplified proplets are held together by a common prn value (here 625). The functor-argument is coded solely in terms of attribute values. For example, the *Julia* and *John* proplets specify their functor as know, while the *know* proplet specifies Julia and John as its arguments.[5]

   A content like 3.2.1 may be turned into a schema by replacing each occurrence of a constant with a variable (simultaneous substitution):

3.2.2 TURNING 3.2.1 INTO A SCHEMA

$$
\begin{bmatrix} \text{noun: } \alpha \\ \text{fnc: } \beta \\ \text{prn: K} \end{bmatrix}
\begin{bmatrix} \text{verb: } \beta \\ \text{arg: } \alpha\ \gamma \\ \text{prn: K} \end{bmatrix}
\begin{bmatrix} \text{noun: } \gamma \\ \text{fnc: } \beta \\ \text{prn: K} \end{bmatrix}
$$

The schema 3.2.2 defines the same semantic relations between *pattern proplets* as does the content 3.2.1 between *content proplets*. A schema matches the content from which it has been derived as well as an open number of similar contents. A DBS schema is not just a *l'art pour l'art* linguistic generalization, but allows using detailed syntactic and semantic properties for efficient high-resolution retrieval (Sect. 6.5). The matching between a schema and a content is illustrated below using the schema 3.2.2 and the content 3.2.1:

---

[5] When we refer to a proplet by its core value, we use italics, e.g., *John*, whereas for reference to an attribute or a value within a proplet, we use helvetica, e.g., fnc or know.

3.2.3 PATTERN MATCHING BETWEEN SCHEMA 3.2.2 AND CONTENT 3.2.1

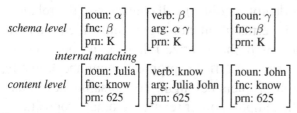

The matching between the schema and the content is successful because the pattern proplets have the same[6] attributes in the same order as the corresponding content proplets and the variables used as values at the schema level match the constants at the content level.

The proplet set of a content is *order-free* in the sense that the storage location of the proplets does not affect the semantic relations defined between them.[7] This is because the semantic relations between proplets are coded by *address* rather than by *embedding* (as in recursive feature structures). For example, the three proplets in 3.2.1 may be represented in the order abc, acb, bac, bca, cab, and cba, and yet maintain their semantic relations intact, as shown below:

3.2.4 MAINTAINING SEMANTIC RELATIONS REGARDLESS OF ORDER

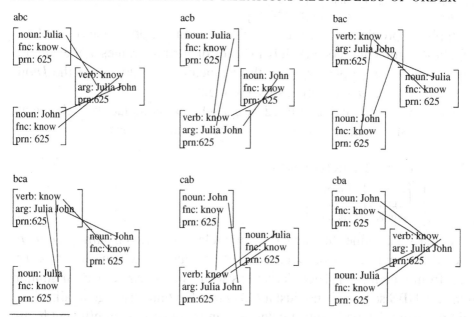

---

[6] As defined in NLC'06, 3.2.3, it is sufficient for successful matching if the attributes of the pattern proplet are a *subset* of the attributes of the content proplet.

[7] This is in contradistinction to a logical formula such as $\exists x[man(x) \wedge walk(x)]$, a phrase structure tree such as 7.1.2, or a recursive feature structure with unification as shown in NLC'06, 3.4.5, which change their meaning or lose their well-formedness if the order of their parts is changed. They also fail to provide a natural primary key (cf. Hausser 2007).

The six possible proplet orders are arranged vertically. The bidirectional semantic relations within each triple are indicated by lines, which are like rubber bands adjusting to the varying arrangements of the proplets. For a better drawing of these lines, the proplet in the middle is moved one step to the right.

The coding method illustrated in 3.2.1–3.2.4 with an example of intrapropositional functor-argument works equally well for extrapropositional coordination (or any other construction of natural language):

3.2.5 Coordination structure of Julia sang. Sue slept. John read.

$$
\begin{bmatrix} \text{noun: Julia} \\ \text{fnc: sing} \\ \text{prn: 10} \end{bmatrix}
\begin{bmatrix} \text{verb: sing} \\ \text{arg: Julia} \\ \text{nc: (sleep 11)} \\ \text{pc:} \\ \text{prn: 10} \end{bmatrix}
\begin{bmatrix} \text{noun: Sue} \\ \text{fnc: sleep} \\ \text{prn: 11} \end{bmatrix}
\begin{bmatrix} \text{verb: sleep} \\ \text{arg: Sue} \\ \text{nc: (read 12)} \\ \text{pc: (sing 10)} \\ \text{prn: 11} \end{bmatrix}
\begin{bmatrix} \text{noun: John} \\ \text{fnc: read} \\ \text{prn: 12} \end{bmatrix}
\begin{bmatrix} \text{verb: read} \\ \text{arg: John} \\ \text{nc:} \\ \text{pc: (sleep 11)} \\ \text{prn: 12} \end{bmatrix}
$$

The propositions with the **prn** values **10**, **11**, and **12** are concatenated by the **pc** (for previous conjunct) and **nc** (for next conjunct) values of the respective verbs. For example, the **nc** value of the second proplet *sing* is **(sleep 11)**, while the **pc** value of the fourth proplet *sleep* is **(sing 10)**.[8]

The proplets in the extrapropositional coordination 3.2.5 are order-free in the same sense as shown in 3.2.4 for the intrapropositional functor-argument 3.2.1. They may likewise be turned into a schema by replacing the constants with variables:

3.2.6 Turning 3.2.5 into a schema

$$
\begin{bmatrix} \text{noun: } \alpha \\ \text{fnc: } \beta \\ \text{prn: K} \end{bmatrix}
\begin{bmatrix} \text{verb: } \beta \\ \text{arg: } \alpha \\ \text{nc: } (\delta\ K{+}1) \\ \text{pc:} \\ \text{prn: K} \end{bmatrix}
\begin{bmatrix} \text{noun: } \gamma \\ \text{fnc: } \delta \\ \text{prn: K{+}1} \end{bmatrix}
\begin{bmatrix} \text{verb: } \delta \\ \text{arg: } \gamma \\ \text{nc: } (\psi\ K{+}2) \\ \text{pc: } (\beta\ K) \\ \text{prn: K{+}1} \end{bmatrix}
\begin{bmatrix} \text{noun: } \phi \\ \text{fnc: } \psi \\ \text{prn: K{+}2} \end{bmatrix}
\begin{bmatrix} \text{verb: } \psi \\ \text{arg: } \phi \\ \text{nc:} \\ \text{pc: } (\delta\ K{+}1) \\ \text{prn: K{+}2} \end{bmatrix}
$$

The schema matches the content 3.2.5 from which it was derived as well as an open number of similar contents. The computational simplicity of the matching procedure depends crucially on the definition of proplets as non-recursive feature structures, with the order of attributes fixed within proplets.

The order-free coding of semantic relations is produced by a time-linear mapping from ordered surfaces to an order-free set of connected proplets in the hear mode (3.3.1, 3.4.1). It is essential for the following functions:

3.2.7 Functions based on the order-free nature of proplets

1. Hear mode: storage of proplets in the content-addressable database of a Word Bank (4.1.1).

---

[8] Extrapropositional address values like (sing 10) are explained in Sect. 4.4.

2. Think mode: selective activation of proplets stored in the Word Bank by means of a navigation along the semantic relations between them, reintroducing a time-linear order (3.3.2, 3.4.2).

3. Speak mode: production of natural language as a time-linear sequence of surfaces based on the selective activation of a navigation (3.3.3, 3.4.3).

4. Query answering: retrieval of content corresponding to a schema (4.2.2).

The mapping from ordered lexical proplets to unordered content proplets in the hear mode and from unordered content proplets to ordered lexical proplets constitute the DBS Conversion Universal Number 3 in 1.1.3.

3.3 Hear, Think, and Speak Modes

This section illustrates the cycle of natural language communication in a user-friendly conceptual format. The format represents the basic units as proplets, but indicates the derivational operations graphically, e.g., by means of arrows. Let us begin with a time-linear surface compositional hear mode derivation:

3.3.1 DBS HEAR MODE DERIVATION OF Julia knows John.

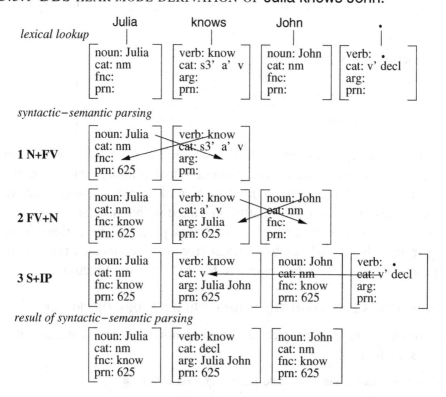

The analysis is surface compositional in that each surface is analyzed as a lexical proplet (lexical lookup, cf. Handl et al. 2009). The derivation is time-linear, as shown by the stair-like addition of one lexical proplet in each new line. Each line represents a derivation step, based on the application of the specified LA-hear grammar rule, e.g., **1 N+FV** (defined in 3.4.1). The rules establish grammatical relations by copying values, as indicated by the diagonal arrows. The result of the derivation is the order-free set of proplets 3.2.1, ready to be stored in the agent's content-addressable memory (4.1.1).

Based on the grammatical relations between the proplets stored in the agent's memory, the second step in the cycle of natural language communication is a *selective activation* of content by navigating from one proplet to the next. The following example is based on the content 3.2.1, derived in 3.3.1:

3.3.2 DBS THINK MODE NAVIGATION

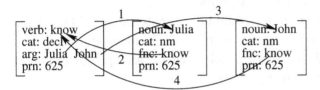

The navigation is driven by an LA-think grammar which uses the grammatical relations between proplets like a railroad system. By constructing proplet addresses from the **arg**, **fnc**, and **prn** values, the navigation proceeds from the verb to the subject noun (arrow 1), back to the verb (arrow 2), to the object noun (arrow 3), and back to the verb (arrow 4).

Such a think mode navigation provides the *what to say* for language production from stored content, while the third step in the cycle of communication, i.e., the speak mode, provides the *how to say it* (McKeown 1985) in the natural language of choice. Consider the following example of a speak mode derivation, resulting in a surface realization:

3.3.3 DBS SPEAK MODE REALIZATION

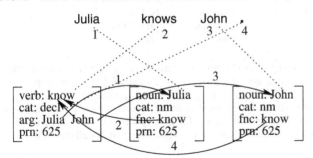

The derivation is based on the same navigation as 3.3.2, wherebby the surfaces are realized from the *goal proplet* of each navigation step, using mainly the core value. In NLC'06, the DBS cycle of communication has been worked out in detail for more than 100 constructions of English.

3.4 Algorithm of LA-Grammar

Having shown informally how the data structure of proplets may be used (i) for mapping surfaces into content (3.3.1), (ii) for activating content selectively (3.3.2), and (iii) for mapping activated content into surfaces (3.3.3) let us turn to the formal rules performing these procedures. Given that the three phases of the cycle of natural language communication are each time-linear, i.e., linear like time and in the direction of time, they may all three be handled by the same algorithm, namely time-linear LA-grammar.[9]

The rules of an LA-hear grammar combine a "sentence start" (*ss*) with a "next word" (*nw*) into a new "sentence start" (*ss'*). Furthermore, the *ss*, the *nw*, and the *ss'* are defined as pattern proplets which are applied to content proplets by means of pattern matching (as in 3.2.3).

Consider, for example, the definition of **N+FV** (for nominative plus finite verb). The rule is shown as it applies to a matching language input, corresponding to the first derivation step of 3.3.1 (explanations in italics):

3.4.1 LA-HEAR RULE APPLICATION

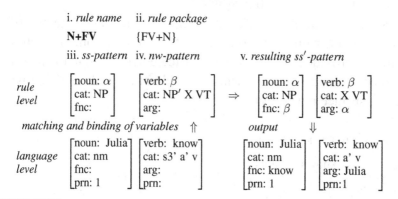

---

[9] LA-grammar is of lower complexity compared to Phrase Structure Grammar regarding relevant language classes (TCS'92). The LA-grammar algorithm for the lowest class C1 has more expressive power than a finite state algorithm (cf. CoL'89, Sect. 8.2). Time-linear applications range from describing the evolution of the universe and the development of an embryo all the way to the control of production lines, the disassembly of old ships, the stages of a soap opera, and organizing a team of workman as to when which task has to be finished for the next tema to appear, for all eventualities.

An LA-hear grammar rule consists of (i) a rule name, (ii) a rule package, (iii) a pattern for an *ss* (sentence start), (iv) a pattern for an *nw* (next word), and (v) a pattern for the resulting (next) sentence start (*ss'*). A pattern at the rule level matches a proplet at the language level if (a) the attributes of the pattern are a subset of the attributes of the proplet and (b) the values of the pattern are compatible with the values of the proplet.

Vertical binding of the variables in the input patterns (rule level) to the corresponding constants in the language input (language level) enables the rule to compute an output. The new content is built by replacing the variables in the output patterns with the constants matching the input part of the rule. The operation is performed at the language level.

For example, by binding the variable α of the input pattern to the corresponding constant *Julia* (language level), the [arg: α] feature of the output pattern (rule level) provides the value Julia to the arg attribute of the output at the language level. If a current rule application is successful, the resulting sentence start is provided with the proplet of a next word, if available, by automatic word form recognition. This results in a new input pair to which the rules of the current rule package are applied.

Variants of the method illustrated in 3.4.1 are used in the other two steps of the communication cycle, i.e., the think mode, based in part[10] on LA-think, and the speak mode, realized by LA-speak. An LA-think grammar powers the navigation from one proplet to the next along the semantic relations between them, resulting in a selective activation of content stored in the Word Bank. An LA-think rule takes a current proplet as input and computes a next proplet as output (successor).

Consider the LA-think rule **VNs** (from verb to noun). It is shown as it executes navigation step 1 from the first to the second proplet in 3.3.2:

3.4.2 LA-THINK RULE APPLICATION

	i. *rule name* **VNs**	ii. *rule package* {NVs}
rule level	iii. *current proplet* $\begin{bmatrix} \text{verb: } \beta \\ \text{arg: X } \alpha \text{ Y} \\ \text{prn: K} \end{bmatrix} \Rightarrow$	iv. *next proplet* $\begin{bmatrix} \text{noun: } \alpha \\ \text{fnc: } \beta \\ \text{prn: K} \end{bmatrix}$
matching and binding of variables	⇑	⇓
Word Bank level	$\begin{bmatrix} \text{verb: know} \\ \text{cat: decl} \\ \text{arg: Julia John} \\ \text{prn: 625} \end{bmatrix}$	$\begin{bmatrix} \text{noun: Julia} \\ \text{cat: nm} \\ \text{fnc: know} \\ \text{prn: 625} \end{bmatrix}$

---

[10] The other task of the think mode is inferencing. Chaps. 5, 6, 10, and 11.

By using the same variables, α, β, and **K**, in the patterns for the current and the next proplet, and by binding them to the values **know**, **Julia**, and **625** of the input proplet *know*, the pattern for the next proplet provides the information required for visiting the successor proplet, here *Julia*.

Finally consider a rule of an LA-speak grammar (see 7.4.3 for a detailed definition). It is like an LA-think rule, except that it is extended to produce appropriate word form surfaces by using the core value as well as the morphosyntactic information of the **cat** and **sem** attributes. The following example shows the LA-speak rule application underlying transition 2 in 3.3.3, which navigates from the noun *Julia* back to the verb *know*, mapping the core value of the goal proplet into the appropriate surface **know+s**.

3.4.3 LA-SPEAK RULE APPLICATION

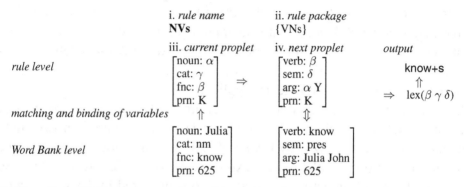

As in an LA-think grammar, the next proplet (here *know*) serves as input for the next rule application(s). The difference between an LA-think and an LA-speak rule is that the latter also produces a surface (here **know+s**), using a variant of the **lex** function defined in NLC'06, 14.3.4. The resulting agent-internal, modality-free surface is used as a blueprint (template) for one of the agent's language synthesis components.

At the most basic level, inter-agent communication in natural language is based on the Second Mechanism of Communication (MoC-2):

3.4.4 THE SECOND MECHANISM OF COMMUNICATION (MoC-2)

The external time-linear surface order is used for coding grammatical relations proplet-internally (hear mode), while the grammatical relations coded proplet-internally are used for coding a time-linear surface order externally (speak mode).

Extensive linguistic research, beginning with NEWCAT'86, has shown that the decoding of grammatical relations from nothing but a time-linear sequence of modality-dependent unanalyzed external surfaces works very well. When the resulting proplets are stored in the Word Bank, any grammatically meaningful order is lost proplet-externally. A time-linear order is (re-)introduced by LA-think, which decodes the proplet-internal addresses to follow the grammatical relations between them. This order is also used by LA-speak.

3.5 Relating Kinds of Proplets to Traditional Parts of Speech

The data structure of proplets is completely general: they may be defined for any set of attributes and any set of values, and still be processed by suitably adapted LA-grammar rules based on pattern matching. However, the long-term upscaling of a talking robot will only be successful if the basic functionality of its artificial cognition is empirically correct – and one window for investigating natural cognition is the analysis of natural language. Let us therefore examine the empirical nature of proplets from a linguistic point of view.

The number and the properties of different kinds of proplets are closely related to the *parts of speech* (PoS) of traditional grammar. We are looking for the minimal part of speech set required for a general computational model of natural language communication.

Despite some considerable variation in the literature on grammar, the traditional parts of speech may be defined roughly as the following eight:

3.5.1 TRADITIONAL PARTS OF SPEECH

1. *verb*
   Includes finite forms like **sang** and non-finite forms like **singing** or **sung** of main verbs, as well as auxiliaries like **was** or **had** and modals like **could** and **should**. Some traditional grammars treat non-finite verb forms as a separate class called *participle*.

2. *noun*
   Includes common nouns like **table** and proper names like **Julia**. Also, count nouns like **book** and mass nouns like **wine** are distinguished.

3. *adjective*
   Includes determiners like **a(n)**, **the**, **some**, **all**, and **my** as well as adnominals like **little**, **black**, and **beautiful**. Some traditional grammars treat determiners as a separate class.

4. *adverb*
   Includes adverbial modifiers like **beautifully** and intensifiers like **very**.

5. *pronoun*

   Includes nouns with an indexical meaning component such as I, me, mine, you, yours, he, him, his, she, her, hers, etc.

6. *preposition*

   Function words which combine with a noun into an adjective, such as on in [the book] on [the table].

7. *conjunction*

   Includes coordinating conjunctions (parataxis) like and and subordinating conjunctions (hypotaxis) like that (introducing subject or object sentence) or when (introducing adverbial sentence).

8. *interjection*

   Includes exclamations like ouch!, greetings like hi!, and answers like yes.

Most traditional grammars postulate eight parts of speech because this is the number adopted by classical Greek and Latin grammars. In daily practice, however, additional classifications are used such as *determiner, auxiliary, modal, infinitive, progressive, past participle, present tense, past tense, singular, plural, first person, second person, third person,* etc., all of which are useful for a more precise classification of word forms.

For linguistic analysis, DBS uses only noun, verb, and adj as core attributes.[11] These have proved sufficient for the DBS analyses[12] of Albanian (Kabashi 2003, 2007), Arabic (ben-Zineb 2010), Bulgarian (Ivanova 2009, Sabeva 2010), Chinese (Mei 2007), Czech, English (Leidner 2000, Proisl 2008, Bauer 2011), French (Pepiuk 2010, Kosche 2011), German (Stefanskaia 2005, Girstl 2006, Gao 2007, Mehlhaff 2007, Tittel 2008, Handl 2010, Weber et al. 2010, Jaegers 2010, Reihl 2010), Georgian, Italian (Weber 2007), Japanese, Korean (Lee 2004, Kim 2009), Polish (Niedobijczuk 2010), Romanian (Pandea 2010), Russian (Vorontsova 2007, Kalender 2010), Spanish (Mahlow 2000, Huezo 2003), Swedish (Lipp 2010), and Tagalog (Söllch 2009).

Using only noun, verb, and adj as core values simplifies the compositional semantics. Also, they are the only parts of speech which have counterparts in logic, namely *argument, functor,* and *modifier,* and in analytic philosophy, namely *object, relation,* and *property* (cf. FoCL'99, Sect. 3.4).

---

[11] Technically, the DBS system allows as many different core attributes as desired.

[12] The languages listed were investigated at the Laboratory for Computational Linguistics at the University of Erlangen Nürnberg (CLUE). In most cases, especially for the more distant languages, the analysis was done by native speakers. In some cases, the analysis did not result in an academic degree or a publication. It did, however, always result in a Malaga (Beutel 2009) or a JSLIM implementation. If the analysis of a language is limited to automatic word form recognition it is nevertheless based on detailed studies of the syntax and semantics.

DBS models the remaining parts of speech by assimilating them into *noun, verb,* and *adjective* proplets. For example, traditional grammar classifies 2. *noun* and 5. *pronoun* as different parts of speech,[13] but in DBS they have the same core attribute noun. This is because a *pronoun* like she can serve the same grammatical function (for example, as subject) as a *proper name* like Julia or a *phrasal noun* like the pretty young girl.

The correct distinction between phrasal nouns, pronouns, and proper names is of a sign-theoretic nature: they use different mechanisms of reference (cf. FoCL'99, Sect. 6.1; NLC'06, Sect. 2.6), namely those of the symbol, the indexical, and the proper name. Given that phrasal nouns and proper names, despite their different reference mechanisms, are traditionally included in the part of speech *noun*, there is no reason to exclude pronouns just because they use yet another mechanism of reference, i.e., indexical. Consequently, in DBS all three kinds of nouns are analyzed as proplets with the same core attribute.

This is illustrated by the following proplets for a common noun, a pronoun, and a proper name. Also included is a determiner because DBS lexically analyzes determiners as proplets with the core attribute noun (7.2.4):

3.5.2 ANALYZING DIFFERENT KINDS OF NOUNS AS LEXICAL PROPLETS

common noun	pronoun	proper name	determiner
sur: books	sur: they	sur: Julia	sur: every
noun: book	noun: ça	noun: Julia	noun: n_1
cat: pn	cat: pnp	cat: nm	cat: snp
sem: count pl	sem: count pl	sem: sg	sem: pl exh
fnc:	fnc:	fnc:	fnc:
mdr:	mdr:	mdr:	mdr:
prn:	prn:	prn:	prn:

The common noun proplet has a core value defined as a concept which is represented as book, serving as a placeholder (6.6.8); analyzed word forms with a concept as their meaning are called *symbols*. The pronoun proplet has a core value defined as a pointer which is represented as ça;[14] analyzed word forms with a pointer as their meaning are called *indexicals*. The *proper name* proplet has a core value defined as a marker which is represented as Julia.

The analysis of determiners (which some traditional grammars treat as a separate part of speech) as noun proplets facilitates the fusion of a determiner and its noun, as illustrated in 7.2.4. The core value of a determiner is a substitution variable; during interpretation in the hear mode, this variable is replaced with the core value of the associated common noun. In some languages, e.g., Ger-

---

[13] This misguided PoS distinction was postulated by Dionysius Thrax (170 – 90 BC).

[14] See Chap. 11 for the interpretation of third person pronouns.

man, determiners are also used indexically, which may be handled by widening the restriction on the substitution variable to include indexical use.

Another difference between the traditional part of speech classification 3.5.1 and the core attributes of DBS concerns 3. *adjective*, 4. *adverb*, and 6. *preposition*. In DBS, they are treated as proplets with the same core attribute, namely adj. Consider the following examples:

3.5.3 ANLYAZING DIFFERENT ADJECTIVES AS LEXICAL PROPLETS

adnominal	*adverbial*	*indexical adjective*	*preposition*
⎡sur: beautiful⎤	⎡sur: beautifully⎤	⎡sur: here ⎤	⎡sur: on ⎤
⎢adj: beautiful⎥	⎢adj: beautiful⎥	⎢adj: idx_loc⎥	⎢adj: *on* n_2⎥
⎢cat: adn ⎥	⎢cat: adv ⎥	⎢cat: adnv ⎥	⎢cat: adnv ⎥
⎢sem: psv ⎥	⎢sem: psv ⎥	⎢sem: ⎥	⎢sem: ⎥
⎢mdd: ⎥	⎢mdd: ⎥	⎢mdd: ⎥	⎢mdd: ⎥
⎣prn: ⎦	⎣prn: ⎦	⎣prn: ⎦	⎣prn: ⎦

While all four proplets share the core attribute adj, they are distinguished in terms of the sign kind of their core values and in terms of their cat values. The proplets beautiful and beautifully share the same core value beautiful (symbol), but differ in their cat values adn (adnominal) and adv (adverbial). The proplets here and on share the cat value adnv (adnominal and adverbial use),[15] but differ in the sign kind of their core values, namely indexical (idx_loc) and preposition plus a substitution variable for a noun (on n_2). All adj proplets have the continuation attribute mdd, for "modified."

Analyzing the adnominal and adverbial uses of an adjective as proplets with the same core value, here beautiful, is partially motivated by terminology: the Latin root of *adjective* means "what is thrown in," which aptly characterizes the optional quality of modifiers in general. It is also motivated morphologically because of the similarity between adnominal and adverbial adjectives. The two uses may resemble each other also in their analytic degrees, as in more beautiful (adnominal) and more beautifully (adverbial) in English.

Languages may differ in whether they treat the adnominal or the adverbial form of an adjective morphologically as the unmarked case. For example, in English the unmarked case is the adnominal,[16] e.g., beautiful, while the adverbial is marked, e.g., beautifully. In German, in contrast, the adnominal

---

[15] As a cat value, adnv is an empirically motivated underspecification in DBS. For a comparison of underspecification in DBS and in Nativism see NLC'06, p. 92, footnote 3.

[16] That the adnominal form happens to equal the unmarked case of adjectives in English may be the reason for the widespread, but nevertheless misguided, terminological practice of calling adnominal use "adjective," in contrast to adverbial use, which is called "adverb." If we were to apply the same logic to German, it would have to be the adverbials which are called "adjective."

Equating adnominal with adjective reduces the number of available terms from three to two. However, three terms are required by the combinatorial possibilities represented by the cat values adnv, adn, and adv.

is marked, e.g., **schöner**, **schöne**, **schönes**, etc., while the adverbial, e.g., **schön**, is unmarked.

The relations between English adjectives with the **cat** values **adnv**, **adn**, and **adv** may be shown as follows (cf. NLC'06, Chap. 15):

3.5.4 RELATION BETWEEN THE adnv, adn, AND adv VALUES IN ENGLISH

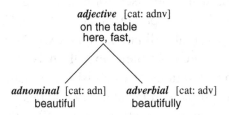

The part of speech is adjective. It can modify nouns as well as verbs, indicated by the **cat** value **adnv**. This value characterizes the combinatorics of prepositional phrases like **on the table**, indexical adjectives like **here**, and certain symbol adjectives like **fast**. If an adjective has two forms, e.g., **beautiful** and **beautifully**, the unmarked form has the **cat** value **adn** (adnominal use) in English, while the marked form has the **cat** value **adv** (adverbial use).

Finally, we turn to the traditional parts of speech which share the core attribute **verb** in DBS. These are finite verbs, auxiliaries, and participles:

3.5.5 ANALYZING DIFFERENT VERB FORMS AS LEXICAL PROPLETS

finite main verb	*finite auxiliary*	*non-finite main verb*
sur: knows	sur: is	sur: knowing
verb: *know*	verb: v_1	verb: *know*
cat: ns3′ a′ v	cat: ns3′ be′ v	cat: a′ be
sem: ind pres	sem: ind pres	sem: prog
mdr:	mdr:	mdr:
arg:	arg:	arg:
prn:	prn:	prn:

Finite verb forms are characterized by a **cat** value ending with the constant **v**. This holds for main verbs and auxiliaries alike. The difference between the two is that auxiliaries, but not main verbs, have the constant **be′**, **hv′**, or **do′** as part of their categories (cf. FoCL9, Sect. 17.3). Also, the core value of an

---

In DBS, "adjective" is the part of speech, which can be used adnominally and adverbially. Therefore "adjective" is used to refer to **adnv** expressions without morphological restrictions, "adnominal" for **adn** expressions morphologically restricted to noun modification, and "adverbial" for **adv** expressions morphologically restricted to verb modification.

In other words, DBS distinguishes between adjectives according to what they modify, regardless of which of the semantic relations happen to be morphologically marked or unmarked in the natural language at hand.

auxiliary (function word) is a substitution value (cf. NLC'06, Sect. 6.4), while that of a main verb is a concept. Non-finite verb forms differ from finite ones by the absence of the constant v at the end of their **cat** value. The verbal moods of the *indicative, subjunctive,* and *imperative* are coded as the **sem** values **ind, sbjv,** and **impv,** respectively (Sect. 5.6; Quirk et al. (1985:149–150)).

There remain the traditional parts of speech *7. conjunction* and *8. interjection.* Conjunctions are function words which are used for connecting expressions of the same part of speech (coordination, parataxis, cf. NLC'06, Chaps. 8 and 9) or of different part of speech (subordination, hypotaxis, NLC'06, Chap. 7). Interjections are one-word sentences; their lexical analysis is not constrained by the compositional considerations of establishing semantic relations and they still await analysis as proplets in Database Semantics.

In summary, despite our using only three core attributes in DBS, the traditional parts of speech may be reconstructed using other properties. For example, based on the lexical analysis of word forms in 3.5.2, *pronouns* as a traditional part of speech are characterized by an indexical core value, while *determiners* have a substitution variable instead. Similarly, based on the lexical analysis in 3.5.3, *prepositions* as a traditional part of speech are characterized by the cat value **adnv** and a substitution variable as their core value, adnominals by the cat value **adn** and a concept as their core value, indexical adverbs by the cat value **adv** and a pointer as their core value, and so on.

These grammatical distinctions may be easily used to lexically define word forms as proplets which have precisely the properties desired: Just invent a nice attribute name and the associated value(s), and put them into your DBS system. The grammatical distinctions are also used in the automatic definition of schemata for equally precise retrieval from the Word Bank. Like all schemata, they constitute a particular *view* on the data (Sect. 4.5).

Finally, another remark on terminology. DBS distinguishes between the three parts of speech **noun, verb,** and **adj,** and between the three levels of complexity *elementary, phrasal,* and *clausal* (Hausser 2009). These are orthogonal to each other and combine as follows:

3.5.6 PARTS OF SPEECH AND LEVELS OF COMPLEXITY

	elementary	*phrasal*	*clausal*
noun	Julia, she	the beautiful girl	that Fido barked
verb	barked	could have barked	Fido barked.
adj	here, beautiful	in the garden	When Fido barked

Accordingly, we say "phrasal noun" when we mean something like **the man in the brown coat** rather than "noun phrase," except in the discussion of the

Bach-Peter sentence, 11.5.9, in which we implicitly adopt the terminology of Transformational Grammar. We also call a word form like man a noun, but this use is on the level of morphology rather than syntax.

3.6 Linguistic Relativism vs. Universal Grammar

The parsing of natural language expressions in the hear mode of DBS results in contents which are partially language-independent. This applies to the following aspects of a content represented as a set of proplets.

First, many content words in different languages, e.g., dog, chien, Hund, and cane, are represented by proplets with the same core value, e.g., the concept *dog*.[17] Second, function words and morphological markings of different languages, for example, singular or past tense, are interpreted using the same values of the cat and sem attributes whenever appropriate. Third, the word order of the surface[18] is dissolved into an order-free set of proplets.

Conversely, DBS content representations are language-dependent in that different languages may use different lexicalizations[19] and different syntactic-semantic constructions[20] to code the same content. This applies not only to languages which are distant from each other, such as English and Chinese, but also to neighboring languages such as English, German, and Italian.

For example, the following sentences have the same meaning₁, but their syntactic-semantic structure in English, German, and Italian is different:

3.6.1 EQUIVALENT CLAUSES WITH DIFFERENT CONSTRUCTIONS

English: I don't care.
German: Es ist mir egal.   (*It is me equal.*)
Italian: Mi lascia indifferente. (*Me leaves-it indifferent.*)

---

[17] Cf. 6.6.3 for the explicit proplet definitions.

[18] For a uniform method of mapping the different word orders in declarative main clauses of German (second position of the finite verb with free order of the obligatory arguments), English (post-nominative position of the verb with fixed order of the obligatory arguments), Korean (final position of the verb with free order of the obligatory arguments), and Russian (free order of the verb and its arguments) into equivalent contents, see Hausser (2008). The order of optional modifiers is usually relatively free.

[19] The literature on lexicalization is vast. See Talmy (1985), Jackendoff (1990), Pustejovsky (1995), Brinton and Traugott (2005), and others. Lexicalization is one area in contemporary linguistics in which the crucial distinction between the literal meaning₁ of a sign and the speaker meaning₂ of an utterance is more or less properly observed. See FoCL'99, 4.3.3, First Principle of Pragmatics.

[20] This notion is central to Construction Grammar (cf. Fillmore et al. 1988, Croft 2000). Constructions are susceptible to a DBS grammatical analysis in terms of parts of speech, functor-argument, and coordination. It is just that they are of a more pronounced collocational and/or idiomatic nature.

The subject position of the English clause may be filled by any non-clausal noun such as **the man with the brown coat** (phrasal) and **you** or **John** (elementary), with the finite verb form suitably adapted. The finite verb form is the auxiliary **do** in its negated form and may take different tense forms. The main predicate is the infinitive of the verb **care**, which could be replaced by the verb **mind** without changing the construction.

The German sentence uses the expletive **es** as the subject.[21] The finite verb is the auxiliary **sein** (corresponding to English **be**) without negation, which may take different tense and mood forms. The German counterpart to the subject of the English clause is any noun in the dative case. The main predicate is the adverbial **egal**, which could be replaced by **gleich** or **wurst** without changing either the form or the meaning of the construction.[22]

And accordingly for Italian. Such asymmetries between different natural languages have long been noted and have fostered a tradition within language theory. It is known as *linguistic relativism* or the *Humboldt-Sapir-Whorf hypothesis*, according to which there does not exist a universal (natural, innate) semantics which all natural languages map into and out of.[23]

Even the part of speech distinctions corresponding to the DBS core attributes **noun, verb,** and **adj** have been found to be absent in some natural languages. For example, *Cayuga*, an Iroquois language, at last count spoken by fewer than one hundred[24] native Americans living in Ontario/Canada, has been argued to have no distinction between verbs and nouns (Sasse 1993), though Rijkhoff (2002) and Hengeveld (1992) disagree.[25]

For the outsider, such controversy is difficult to assess. First, for one not speaking Cayuga natively, it is most precarious to judge the *empirical* facts; they concern not only the surfaces provided by the tape recorder, but also the intuitions of the speaker-hearer. Second, there is the *theoretical* question of whether the claim applies to morphology, syntax-semantics, or pragmatics.

As a related phenomenon consider *conversion* in English: a surface like **make** can be used as a noun in **Jim collects all makes** and as a verb in **Jim makes toys.**[26] By analogy, the claimed absence of nouns in Cayuga may

---

[21] Grammatically acceptable subject alternatives such as **der Preis** would compromise the equivalence with the English counterpart.

[22] Intensified versions of **egal** in German are **schnurzegal** and **schnurzpiepegal**. The **piep** in the latter has been suggested as stand-in for a swearword.

[23] For a modern argument from language typology, see J. Nichols (1992).

[24] Even if a language has only a few speakers left and therefore little commercial promise, a DBS implementation would be worthwhile for the comparison with other languages. It would also serve research in typology, anthropology, psychology, sociology, and so on.

[25] Mithun's (1999) formidable overview of the Iroquois languages lists the paper by Sasse in the bibliography, but does not discuss or mention it in the text.

apply to (i) nouns and verbs sharing a stem, like **make** in English, or (ii) nouns and verbs sharing a stem and morphological alternations, like **make/make+s** in English, or it may mean (iii) that the native speakers of Cayuga do not distinguish between objects and relations. Position (iii) corresponds to the radical linguistic relativism of Whorf (1964).

---

[26] The DBS system for English treats the concept *make* as a core value which may be embedded into the proplet shell of a noun or a verb (6.6.5 – 6.6.6 for related examples).

4. Mystery Number Three: Memory Structure

Because the relations between proplets within a proposition and between propositions are coded solely by means of proplet-internal attribute values (addresses), proplets are inherently *order-free*. Thus proplets can be stored in accordance with any kind of database schema – without losing their inter-proplet relations, but gaining the use of the storage and retrieval mechanism of the database selected. Given that there are several different database schemata available, the question of the third mystery is: which one is suited best for modeling cognition, including the cycle of natural language communication?

4.1 Database Schema of a Word Bank

When faced with the choice of a computer memory, the most basic alternative is between a *coordinate*-addressable and a *content*-addressable approach (cf. Chisvin and Duckworth 1992 for an overview). Though peppered with patents, the content-addressable approach is less widely used than the coordinate-addressable approach. A content-addressable memory is suited best for the super-fast retrieval of content which is written once and never changed.

A coordinate-addressable memory, e.g., an RDBMS, resembles a modern public library in which a book can be stored wherever there is space (random access) and retrieved using a separate index (inverted file) relating a primary key (e.g., author, title, year) to its location of storage (e.g., 1365). A content-addressable memory, in contrast, is like a private library in which books with certain properties are grouped together on certain shelves, ready to be browsed without the help of a separate index. For example, at Oxford University the 2 500 volumes of Sir Thomas Bodley's library from the year 1598 are still organized according to the century and the country of their origin.

In an initial response to a content-addressable approach, mainstream database scientists pointed out that it can be simulated by the coordinate-addressable approach (Fischer 2002), using well-established relational databases. The issue here, however, is whether or not the formal intuitions of the content-addressable approach can be refined naturally into a model of cognition.

Our point of departure is the data structure of proplets. For purposes of storage and retrieval, a proplet is specified uniquely[1] by its **core** and **prn** values (primary key). This suggests a two-dimensional database schema, as in a classic network database (cf. Elmasri and Navathe 1989). A column of owner records is in the alphabetical order of their core values. Each owner record is preceded by a list of member records, distinguished between in terms of their **prn** values. However, instead of using member and owner *records* we use equivalent member and owner *proplets*. The result is called a Word Bank.

As an example, consider storing the proplets of the content 3.2.1:

4.1.1 STORING THE PROPLETS OF 3.2.1 IN A WORD BANK

	member proplets	*now front*	*owner proplets*
		...	...

$$
\cdots\ \begin{bmatrix} \text{noun: John} \\ \text{cat: nm} \\ \text{fnc: ...} \\ \text{prn: 610} \end{bmatrix} \begin{bmatrix} \text{noun: John} \\ \text{cat: nm} \\ \text{fnc: know} \\ \text{prn: 625} \end{bmatrix} \qquad [\text{core: John}]
$$

$$
\cdots\ \begin{bmatrix} \text{noun: Julia} \\ \text{cat: nm} \\ \text{fnc: ...} \\ \text{prn: 605} \end{bmatrix} \begin{bmatrix} \text{noun: Julia} \\ \text{cat: nm} \\ \text{fnc: know} \\ \text{prn: 625} \end{bmatrix} \qquad [\text{core: Julia}]
$$

$$
\cdots\ \begin{bmatrix} \text{verb: know} \\ \text{cat: decl} \\ \text{arg: ...} \\ \text{prn: 608} \end{bmatrix} \begin{bmatrix} \text{verb: know} \\ \text{cat: decl} \\ \text{arg: Julia John} \\ \text{prn: 625} \end{bmatrix} \qquad [\text{core: know}]
$$

An owner proplet and the preceding member proplets form a *token line*. The proplets in a token line all have the same core value and are in the temporal[2] order of their arrival, reflected by their **prn** values.

In contrast to the task of designing a practical schema for arranging the books in a private library, the sorting of proplets into a Word Bank is simple and mechanical. The letter sequence of a proplet's core value completely determines its token line for storage: the storage location for any new arrival is the penultimate position in the corresponding token line, called the *now front*.

A Word Bank is content-addressable because no separate index (inverted file) is required. Furthermore, a Word Bank is scalable (a property absent or problematic in some other content-addressable systems). The cost of insertion is constant, independent of the size of the stored data, and the cost of retrieving a specified proplet grows only logarithmically with the data size (external access) or is constant (internal access). External access to a proplet requires(i)

---

[1] Propositions containing two or more proplets with the same values, as in **Suzy loves Suzy**, require extra attention. They constitute a special case which (i) occurs very rarely and (ii) is disregarded here because it is easily handled.

its core and (ii) its **prn** value, e.g., **know 625**. Most cognitive operations, however, require internal access based on addresses (pointers; cf. Sect. 4.4).

Compared to the classic 1969 CODASYL network database, a Word Bank is highly constrained. First, the member proplets belonging to an owner proplet are listed in the temporal order of their arrival. Second, the members in a token line must share the owner's core value (no multiple owners). Third, the only connections between proplets across token lines are the semantic relations of functor-argument and coordination. Fourth, like the relations between owners and members, the semantic connections are 1:n relations: one functor – several possible arguments; one first conjunct – several possible successors; one original – several possible coreferent address proplets.

A Word Bank is a kind of navigational database because it supports the navigation from one proplet to the next, using the semantic relations between proplets (3.3.2) and along token lines (4.2.2) like a railroad system, with the algorithm of LA-grammar (3.4.2) as the locomotive. However, while the navigational databases of the past (Bachman 1973) and the present (XPath, Kay 2004) are intended to be driven by external human users, the system presented here is located inside an artificial cognitive agent, serving as the container and structural foundation of autonomous control (Chap. 5).

4.2 Retrieving Answers to Questions

Because the proplets derived in the hear mode (3.3.1) have a **core** and a **prn** value, they are suitable for (i) storage in a Word Bank (4.1.1). For the same reason, stored proplets support the operation of (ii) navigating from a given proplet to a successor proplet across token lines (3.3.2) in one of the two basic kinds of think mode. Moreover, because there is a speak mode which is riding piggyback on the think mode (3.3.3), the proplets in a Word Bank are suitable (iii) for language production from stored content as well.

Another operation enabled by proplets in a Word Bank is (iv) retrieving answers to questions. This operation is based on moving a query pattern along a token line until matching between the pattern and a proplet is successful. A query pattern is defined as a proplet with at least one variable as a value.

Consider an agent thinking about girls. This means activating the corresponding token line, as in the following example:

---

[2] The token line for any core value is found by using a trie structure (Fredkin 1960). The search for a proplet within a token line may use the **prn** value of the address in relation to the strictly linear increasing **prn** values. Technically, this may be based on binary search, in time $O(log(n))$ (Cormen et al. 2009), or interpolation, in time $O(log(log(n)))$ (Weiss 2005), where n is the length of the token line.

4.2.1 EXAMPLE OF A TOKEN LINE

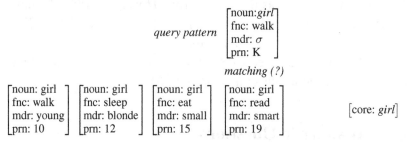

As indicated by the **fnc** and **mdr** values of the member proplets, the agent happened to observe or hear about a young girl walking, a blonde girl sleeping, a small girl eating, and a smart girl reading.

For retrieval, the member proplets of a token line may be checked systematically by using a pattern proplet as the query. The following example shows the pattern proplet representing the query **Which girl walked?** as it applies to the token line 4.2.1:

4.2.2 APPLYING A QUERY PATTERN

$$
query\ pattern \quad
\begin{bmatrix}
\text{noun:} girl \\
\text{fnc: walk} \\
\text{mdr: } \sigma \\
\text{prn: K}
\end{bmatrix}
$$

matching (?)

$$
\begin{bmatrix}
\text{noun: girl} \\
\text{fnc: walk} \\
\text{mdr: young} \\
\text{prn: 10}
\end{bmatrix}
\begin{bmatrix}
\text{noun: girl} \\
\text{fnc: sleep} \\
\text{mdr: blonde} \\
\text{prn: 12}
\end{bmatrix}
\begin{bmatrix}
\text{noun: girl} \\
\text{fnc: eat} \\
\text{mdr: small} \\
\text{prn: 15}
\end{bmatrix}
\begin{bmatrix}
\text{noun: girl} \\
\text{fnc: read} \\
\text{mdr: smart} \\
\text{prn: 19}
\end{bmatrix}
\qquad
\begin{bmatrix}
\text{core: } girl
\end{bmatrix}
$$

The indicated attempt at matching (?) fails because the **fnc** values of the pattern proplet (i.e., **walk**) and of the member proplet (i.e., **read**) are incompatible. The same holds after moving the pattern proplet one member proplet to the left. Only after reaching the leftmost member proplet is the matching successful. Now the variable σ is bound to the value **young** and the variable **K** to the value **10**. Accordingly, the answer provided to the question **Which girl walked?** is **The young girl (walked)**.[3] A powerful extension of this method is combining pattern proplets into schemata (3.2.1–3.2.6, 6.5.1)

4.3 Reference as a Purely Cognitive Procedure

In analytic philosophy of language and in linguistics, the notion of *reference* is generally defined as a "relation between language and the world." The currently most widely used reconstruction of reference is Model Theory, which defines the world as a set-theoretic model (as in Montague grammar; cf. FoCL'99, Sects. 19.3 and 20.4):

---

[3] For a more detailed presentation including **yes/no** questions, see NLC'06, Sect. 5.1.

4.3.1 MODEL-THEORETIC RECONSTRUCTION OF REFERENCE

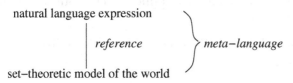

The model, the language analysis, and the reference relation between the expressions and the model are defined in a metalanguage – without any cognitive agents and therefore without any agent-internal database. Furthermore, a computer does not understand the logicians' metalanguage. Instead, it requires the definition of procedural operations (cf. FoCL'99, Sect. 19.4; Schnelle 1988).

To reconstruct reference as a computational procedure, independent of any metalanguage, the DBS robot is based on the following component structure:[4]

4.3.2 INTERFACES AND COMPONENTS OF AN AGENT WITH LANGUAGE

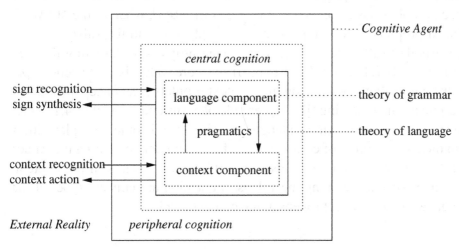

The agent is represented as a body in the real world with interfaces for recognition and action. These provide the link between the agent-external reality and the agent-internal cognition. Cognition is in large part a memory structure which is divided into a language component for processing language data and a context component for processing nonlanguage data.[5]

The language and the context component code content the same way, namely as proplets in a Word Bank, concatenated by the relations of functor-argument

---

[4] Borrowed from NLC'06, 2.4.1. See 4.5.3 and 4.5.4 for a refined version.

[5] The distinction between the theory of language and of grammar is emphasized in Lieb (1976).

and coordination. This similarity of format allows us to base the interaction between the two components on yet another application of pattern matching:[6]

4.3.3 REFERENCE AS LANGUAGE-CONTEXT PATTERN MATCHING

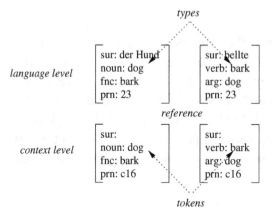

Language and context proplets differ in that the **sur** (for surface) attribute of language proplets has a non-NIL value, e.g., **der Hund**, whereas the **sur** value of context proplets is NIL, represented as empty space in the value slot, i.e., [**sur:**]. Another difference is that in language proplets representing the sign kind of symbols (cf. FoCL'99, Chap. 6) the core value is a concept[7] *type*, while the core value of corresponding context proplets is a concept *token*.

In summary, the matching (i) between rule patterns and content (3.4.1, 3.4.2, 3.4.3) and (ii) between (symbol) language proplets and context proplets differ only in the compatibility of certain values. In rule-content matching the values in question are restricted variables in the rule and constants in the content. In language-context matching the values in question are concept types at the language level and concept tokens at the context level.

4.4 Coreference-by-Address

Computer science uses the term "reference" differently from analytic philosophy and linguistics. A computational reference is the address of a storage location. This may be coded (i) as a symbolic address (declarative) or (ii) as a pointer to the physical storage location in the memory hardware (procedural).

---

[6] Neither the formulas of typed lambda calculus in SCG'84 (Montague Grammar) nor the trees generated from frames in CoL'89 (FrameKit, Carbonell and Josef 1986) proved to be practical for implementing pattern matching between a language component and a context component. For a summary, see FoCL'99, Sect. 22.2.

[7] These concepts provide the place for integrating the prototype theory of Rosch (1999) into DBS.

By accessing data items directly by storage location, pointers are faster than symbolic addresses. Pointers provide instant access in constant time, and are especially suited for data which are written once and never changed – as in the content-addressable memory of DBS.

Because pointers to a physical storage location do not lend themselves to a declarative specification, our representation uses symbolic addresses instead. For example, during language interpretation, symbolic addresses are established by means of copying. The following example shows the first derivation step of 3.3.1, with the result of cross-copying, but before next word lookup:

4.4.1 SYMBOLIC ADDRESSES RESULTING FROM COPYING

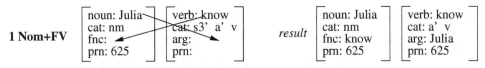

The symbolic addresses resulting from the indicated cross-copying are (i) (know 625)[8] in the *Julia* proplet and (ii) (Julia 625) in the *know* proplet.

When the proplet *Julia* in 4.4.1 is activated in the course of a navigation, the symbolic address (know 625) allows the navigation to continue to the relevant *know* proplet via external access: go to the owner proplet know and look for the member proplet with the prn value 625 (4.1.1).

Computationally, however, internal access is more efficient. It is implemented as a pointer from the fnc attribute of the *Julia* proplet to the physical storage location of the relevant *know* proplet. While the symbolic copying operations are from core values to continuation attributes, the corresponding pointers are in the opposite direction, as needed for navigation (3.3.2).

The coding of inter-proplet relations by means of addresses which are (i) specified declaratively and (ii) implemented as pointers combines a declarative specification with a different, but functionally equivalent, computational realization of high efficiency. The symbolic addresses are also a practical necessity because they allow us to recompute the physical storage locations after occasional clean-ups of the agent's memory (Sect. 5.6).

Another use of addresses, symbolic and by pointer, is for connecting new content to old content by means of coreference. Consider, for example, a cognitive agent observing at moment t_i that *Julia is asleep* and at t_j that *Julia is*

---

[8] This symbolic address is constructed from the attributes fnc: know and prn: 625 in the resulting first proplet. In *intra*propositional relations, such as those in 4.4.1, it would be redundant to specify a prn value, e.g., 625, in the attribute storing the symbolic address, e.g., fnc: (know 625), because it equals the prn value of the proplet containing the attribute. However, if a symbolic address refers to a proplet of another proposition, as in an *extra*propositional coordination (3.2.5), the prn value of the goal proplet must be specified in the address, e.g., [nc: (read 12)].

awake, referring to the same person. Instead of representing this change by revising the first proposition into the second,[9] the second proposition is added as new content, like sediment, leaving the first proposition unaltered:

4.4.2 COREFERENTIAL COORDINATION IN A WORD BANK

$$
\ldots \begin{bmatrix} \text{noun: Julia} \\ \text{fnc: sleep} \\ \text{prn: 675} \end{bmatrix} \ldots \quad \begin{bmatrix} \text{noun: (Julia 675)} \\ \text{fnc: wake} \\ \text{prn: 702} \end{bmatrix} \ldots [\text{core: Julia}]
$$

$$
\ldots \qquad\qquad\qquad \ldots \begin{bmatrix} \text{verb: wake} \\ \text{arg: (Julia 675)} \\ \text{prn: 702} \end{bmatrix} \ldots [\text{core: wake}]
$$

$$
\ldots \begin{bmatrix} \text{verb: sleep} \\ \text{arg: Julia} \\ \text{prn: 675} \end{bmatrix} \ldots \qquad\qquad \ldots [\text{core: sleep}]
$$

The core attribute of the **Julia** proplet with the **prn** value **702** has the address value [**noun: (Julia 675)**], instead of a regular core value, e.g., [**noun: Julia**].[10]

This method, called coreference-by-address, enables a given item to code as many relations to other proplets as needed, without increasing the number of attributes or values within proplets. For example, the proplets in the token line of *Julia* in 4.4.2 have the **fnc** value **sleep** in proposition 675, but **wake** in proposition 702. The most recent (and thus most up-to-date) content relating to the original proplet is found by searching the relevant token line from right to left, i.e., in the anti-temporal direction.

Coreference-by-address establishes a relation between the core values of two proplets,[11] in contradistinction to functor-argument and coordination, which establish a relation between the core value of one proplet and the continuation value of another. In other words, coreference-by-address relates proplets within the same token line, while functor-argument and coordination relate proplets across different token lines.

Defining core values as addresses does not affect the strictly time-linear order of storing proplets in a token line (4.4.2), yet allows us to refer back to a coreferent proplet within that token line. As a result, coreference-by-address provides for a third kind of LA-think navigation – in addition to moving along

---

[9] A more application-oriented example would be *fuel level high* at t_i and *fuel level low* at t_j.

[10] In the JSLIM implementation of this particular use of an address, the relation is realized not only by a pointer to the original *Julia* proplet with the **prn** value **675**, but also by a second, bidirectional pointer (not shown) which connects the current proplet to its immediate coreferential predecessor, resulting in an incremental chain to the original, assuming that there are several coreferent address proplets.

[11] In contrast to coreferential pronouns in natural language, computational coreference-by-address is not limited to nouns. For example, in the sequence **Fido barked. ... When Fido barked.**, the second **barked** may be represented as coreferent with the first.

the semantic relations across token lines (3.3.2), and moving a query pattern along a token line (4.2.2). Consider the following example:

4.4.3 COREFERENTIAL NAVIGATION

$$
\begin{bmatrix} \text{verb: sleep} \\ \text{arg: Julia} \\ \text{prn: 675} \end{bmatrix} \overset{1}{\leftrightarrow} \begin{bmatrix} \text{noun: Julia} \\ \text{fnc: sleep} \\ \text{prn: 675} \end{bmatrix} \overset{2}{\leftrightarrow} \begin{bmatrix} \text{noun: (Julia 675)} \\ \text{fnc: wake} \\ \text{prn: 702} \end{bmatrix} \overset{3}{\leftrightarrow} \begin{bmatrix} \text{verb: wake} \\ \text{arg: (Julia 675)} \\ \text{prn: 702} \end{bmatrix}
$$

Connections 1 and 3 are intrapropositional functor-argument relations between (order-free) *Julia* and *sleep* and *Julia* and *wake*, respectively. Connection 2 is extrapropositional and based on the coreference between the address proplet of proposition 702 and the original *Julia* proplet of proposition 675. The content of 4.4.3 may be realized in English as Julia was asleep. Now she is awake (see Chap. 11 for the coreferential interpretation of third person pronouns).

4.5 Component Structure and Functional Flow

In DBS, pattern matching based on the type-token distinction[12] is used for the following applications:

4.5.1 PATTERN MATCHING BASED ON THE TYPE-TOKEN RELATION

a. *Recognition*:
matching between concept types and raw input (cf. NLC'06, Sect. 4.3)

b. *Action*:
matching between concept tokens and concept types (cf. NLC'06, Sect. 4.4)

c. *Reference*:
matching between language and context proplets (cf. 4.3.3; 6.4.3; NLC'06, 3.2.4; FoCL'99, Sect. 4.2)

Pattern matching based on restricted variables, in contrast, is used to establish the relation between an LA-grammar rule and a Word Bank content:

4.5.2 PATTERN MATCHING BASED ON RESTRICTED VARIABLES

a. *Natural Language Interpretation*:
matching between LA-hear rules and language proplets (3.4.1)

---

[12] The type-token distinction was introduced by C. S. Peirce (1933), *Collected Papers*, Vol.4:537. Steels (1999) showed that new types may be derived automatically from similar data by abstracting from what they take to be variable, and therefore accidental. See also NLC'06, Sect. 4.2, and FoCL'99, Sect. 3.3.

b. *Navigation*:
   matching between LA-think rules and content proplets (3.4.2)
c. *Production from Stored Content*:
   matching between LA-speak rules and content proplets (3.4.3)
d. *Querying*:
   matching between query patterns and content proplets (4.2.2)
e. *Inferencing*:
   matching between inference rules and content proplets (5.2.3 and 5.3.4)

How should the kinds of DBS pattern matching 4.5.1 and 4.5.2 be integrated into the component structure and functional flow of a cognitive agent?

The component structure of diagram 4.3.2 models reference as viewed in analytic philosophy, i.e., as a vertical relation between horizontal language expressions and a horizontal world (4.3.1). At the same time, 4.3.2 departs from the standard assumptions of analytic philosophy, including Truth-Conditional Semantics, because it does not treat reference as an external relation defined in a metalanguage, as postulated by mathematical realism (cf. FoCL'99, Chaps. 19–21), but as an agent-internal, cognitive procedure.

Diagram 4.3.2 is essential for explaining the Seven Principles of Pragmatics in the SLIM theory of language (cf. NLC'06, Sect. 2.6.) and well-suited for showing the applications of pattern matching based on the type-token relation (4.5.1). It fails, however, to provide a place for pattern matching based on restricted variables (4.5.2, a–e). Consider the following alternative component structure, which is functionally more inclusive than diagram 4.3.2:

4.5.3 REFINED COMPONENT STRUCTURE OF A COGNITIVE AGENT

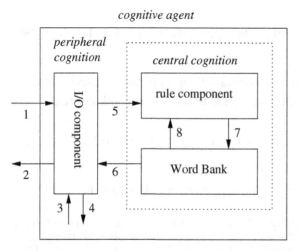

1 = external recognition
2 = external action
3 = internal recognition
4 = internal action
5 = input to rule component
6 = output of Word Bank
7 = rule-Word_Bank interaction
8 = Word_Bank-rule interaction

A general I/O component for external and internal recognition and action unifies the input-output channels for the language and the context level, shown separately in diagram 4.3.2. There is a rule component and a content component (Word Bank), which provide the structural basis for the rule level to govern the processing of content at the memory level (7) – with data-driven feedback from the memory level to the rule level (8).

The rule component and the Word Bank are each connected unidirectionally to the I/O component. All recognition output of the I/O component is input to the rule component (5), where it is processed and passed on to the Word Bank (7). All action input to the I/O component comes from the Word Bank (6), derived in frequent (8, 7) interactions with the rule component.

The operations of the I/O component are realized by the pattern matching 4.5.1 (a) and (b), which are based on the type-token relation. The interaction between the rule component and the Word Bank is realized by the pattern matching operations 4.5.2 (a–e), which are based on restricted variables. The only application of pattern matching listed in 4.5.1 and 4.5.2 which apparently has no place in 4.5.3 is 4.5.1 (c), i.e., reference. This is because it is entirely contained in the Word Bank of 4.5.3, as shown below:

4.5.4 INTEGRATING DIAGRAM 4.3.2 INTO DIAGRAM 4.5.3

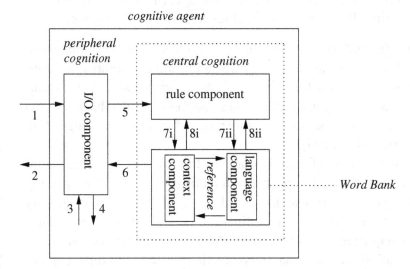

The separate external interfaces for the language and the context component in diagram 4.3.2 are recreated in 4.5.4 by dividing 7 and 8 of 4.5.3 into 7i and 8i for the context level, and into 7ii and 8ii for the language level.

While 4.5.4 serves to reconciliate the different component structures of 4.3.2 and 4.5.3 conceptually, it is misleading insofar as the Word Bank actually contains nothing but owner proplets and member proplets organized in time-linear token lines. It is just that the proplets stored in a Word Bank support several different *views*, among them those of diagrams 4.3.2 and 4.5.3.

Different views are created by referring to different kinds of proplet values: proplets with the same **prn** value are viewed as propositions, proplets with non-NIL **sur** values as language content, proplets with NIL **sur** values as context content, proplets with interlocking **pc** and **nc** values as coordinations, proplet sets with successive **prn** values as text, etc. (1.3.2, 6.6.1–6.6.5).

Therefore the technically correct way of integrating diagram 4.3.2 into diagram 4.5.3 is by changing from a naive-conceptual view to a database view: instead of viewing content proplets as sets with a common **prn** value (propositions), separated into a language and a context level (4.3.2), and so on,[13] the same proplets are viewed solely as items to be sorted into token lines according to their core values and in the order of their arrival.[14]

This conclusion is formulated as the Third Mechanism of Communication:

4.5.5 THE THIRD MECHANISM OF COMMUNICATION (MOC-3)

> The operations of cognition in general and of natural language communication in particular require a memory with a storage and retrieval mechanism supporting (i) extensive data coverage, (ii) functional completeness, and (iii) efficiency which enables real-time performance.

In DBS, extensive (i) data coverage is achieved by allowing an unlimited number of proplet values. These are used to define an unlimited number of views which are the basis of (ii) functional completeness. Performance in (iii) real-time is based on combining a simple content-addressable database schema (Word Bank), a simple data structure (proplets), and a time-linear algorithm running in linear time (C1-LAG).

The need for completeness of data coverage is illustrated by automatic word form recognition (Sect. 2.5). It may be systematically improved by parsing corpora to find missing entries to be added to the online lexicon, and by fine-tuning the rules for inflection/agglutination, derivation, and compounding. Other areas requiring completeness of data coverage are the syntactic-semantic interpretation and the semantic-syntactic production of different language constructions, as well as nonlanguage and language inferencing.

Functional completeness has been illustrated by the operations of (a) storing content derived in the hear mode (3.3.1, 4.3.1), (b) selectively activating content in the think mode (3.3.2), (c) realizing natural language surfaces from

---

[13] In DBS, these different views are established by patterns, and not by *create view*, as in SQL.

[14] See 6.4.3 for an example showing reference as a "horizontal" relation within a token line.

stored content in the speak mode (3.3.3), and (d) finding content corresponding to a question (Sect. 4.2). Other requirements on functional completeness
are autonomous control and inferencing (Chap. 5), as well as adaptation and
learning (Chap. 6).

Computational efficiency has been addressed by showing that the storage of
new proplets in a Word Bank and the retrieval (visiting) of proplets via pointers
requires only constant time (Sects. 4.1 and 4.4). Also, earlier work (TCS'92)
has shown that the applications of LA-grammar to natural language parse in
linear time (linear complexity). All further refinements of the DBS system will
have to be shown not to jeopardize these important results.

4.6 Embedding the Cycle of Communication into the Agent

The component structure 4.5.3 raises the question of how to integrate the LA-
hear, LA-think, and LA-speak derivations outlined in Sects. 3.3 and 3.4 step-
by-step into the functional flow. Furthermore, what are the impulses initiating
these procedures, and where do these impulses come from?

Within the component structure and functional flow of 4.5.3, the hear mode
derivation 3.3.1 may be shown as follows (using the same numbering as in
4.5.3 to indicate corresponding inter-component mappings):

4.6.1 MAPPING INCOMING SURFACES INTO CONTENT (HEAR MODE)

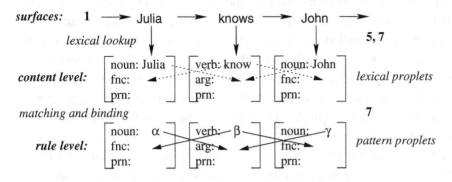

The impulse activating the hear mode operations are the surfaces (1) which the
I/O component provides to the rule component (5), triggering lexical lookup
in the Word Bank (7). As a result, lexical proplets are added one by one at
the end of the corresponding token lines (*now front*) of the Word Bank. At the
same time, the LA-hear grammar rules (3.4.1) of the rule component connect

these lexical proplets incrementally via copying (7) in a strictly time-linear derivation order.

Next consider the language production from stored content 3.3.3 within the component structure of diagram 4.5.3:

4.6.2 MAPPING STORED CONTENT INTO SURFACES (SPEAK MODE)

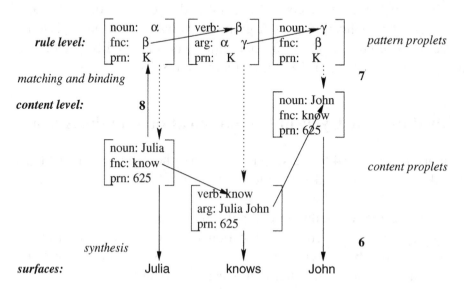

The impulse (8) may be provided by another agent's question (Sect. 4.2) or by a request to recount a certain event.[15] The navigation through the content is driven by the rule level and transmitted to the content level by means of pattern matching (7).[16] Once the LA-think grammar has been started by an initial impulse matching its start state, the derivation is powered by successful rule applications calling successor rules with their rule package. The blueprints for well-formed language expressions passed to the I/O component (6) are derived in frequent (7, 8) interactions between the rule and the memory component.

Finally consider language production based on the on-the-fly inferencing of the agent's autonomous control. Anticipating the discussion of inferences in Chaps. 5, 6, 10, and 11, it is sufficient for present purposes that the output of a DBS inference be a sequence of proplets which (i) have core and prn values and (ii) code the semantic relations of functor-argument, coordination, and coreference.

In other words, the content produced by on-the-fly inferencing has the same format as stored content resulting from language recognition – and can there-

---

[15] For a general discussion of statement, question, and request dialogues, see Chaps. 10 and 11.
[16] For simplicity, the navigation is more direct than in 3.3.3.

fore be processed by the same language-dependent LA-speak rules for language production. As an example, consider the inference β *hungry* **cm** β *eat food* (discussed in Sects. 5.1, 5.3, and 5.6). The connective **cm** stands for countermeasure:

4.6.3 INFERENCE PRODUCING OUTGOING SURFACES

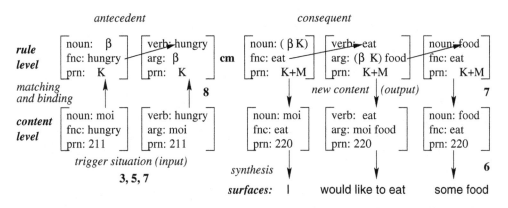

The impulse activating this inference is a sensation of hunger provided by the agent's I/O component (antecedents 3, 5, 7, 8). One way to use the content derived by the consequent (7) would be as a blueprint for nonlanguage action. In 4.6.3, however, the newly derived content is mapped into language, utilizing the core values for the synthesis of surfaces (6). This kind of language production is used especially in dialogue (Chap. 10), which requires and triggers the agent's reasoning for real-time reactions to the actions of the partner in communication.

5. Mystery Number Four: Autonomous Control

The fourth mystery is how to design autonomous control as an upscaling from the cycle of communication by utilizing the conceptual and computational constructs of DBS described so far. The overall goal of autonomous control is to enable survival in the agent's ecological niche. For this, control must guide behavior by connecting the agent's recognition to meaningful action, including language interpretation (recognition) and production (action).

5.1 Pinball Machine Model of Cognition

According to Bernard (1865) and Wiener (1948), the task of autonomous control is to maintain the agent in a continuous state of balance (equilibrium, homeostasis)[1] vis à vis constantly changing external and internal environments. DBS utilizes this task as the main motor driving the agent's cognitive machine. To maintain balance, the agent's cognition continuously produces blueprints for appropriate actions, with non-action as a limiting case.

It follows that a large part of autonomous control is retrieval. Blueprints for action are derived by relating current content to past experiences stored in the agent's memory. The search for suitable action models consists of (i) retrieving contents which match the agent's current concern to various degrees and at various levels of abstraction, and (ii) evaluating their outcome.

The better the available models fit the current task, the less the agent must rely on trial and error. This applies also to behavior not directly related to survival: human desires such as those for power, love, belonging, freedom, and fun may be subsumed under the balance principle by treating them as part of the agent's internal environment – like hunger.

While making relevant contents available in real-time is a necessary condition for maintaining the agent's balance, it is not sufficient. This may be shown by the following comparison. The books in a private library and the proplets in

---

[1] In more recent research, Herrmann (2003) uses the balance principle as the difference between the agent's *ist-Zustand* (as-is-state) and *soll-Zustand* (as-should-be-state) to trigger language production.

the content-addressable memory of a Word Bank (Sect. 4.1) have in common that they are being browsed by a cognitive agent. It is just that in a private library the browsing agent is located inside the arrangement of books on their shelves, while the proplets in a Word Bank are located inside the cognitive agent. As a consequence, the role of the browsing user in a private library seems to be unfilled in a Word Bank.

DBS fills this role, however, with a focus which is like a point of light navigating from one proplet to the next. Unlike with the browsing user in a private library, the control is not located in the focus. Instead, the motion of the focus is controlled in part by the semantic relations between the proplets, which are used like a railroad system for the navigation.[2] The other part is provided by LA-think rules and inferences which relate the agent's current situation to precedents in memory and derive new content serving as blueprints for actions to maintain the agent in a state of balance.

Putting it metaphorically, the navigation of the focus through the railroad system of connected proplets resembles a pinball machine (cf. L&I'10), with the focus serving as the ball and the slanted playing field with its springs and levers serving as the railroad system. Because cognitive operations are agent-internal, there is no "retrieval" in the usual sense. Instead, the navigation of the focus merely *visits* successor proplets – as an activation which may be visualized as lighting up the proplets traversed, with an afterglow of suitable duration and rate of decay (annealing).

The goal of maintaining the agent in a state of balance provides a computational reconstruction of *intention*, just as the agent's recognition and action procedures provide a computational reconstruction of *concepts*, which DBS reuses as the *meanings* of language. This differs from Grice (1957, 1965, 1969), who uses an undefined notion of intention to define natural language meaning:

5.1.1 DEFINITION OF MEANING BY GRICE

Definiendum: U meant something by uttering x.
Definiens: For some audience A, U intends his utterance of x to produce in A some effect (response) E, by means of A's recognition of the intention.

This charming definition uses "intends" and "recognition of the intention" in the definiens as intuitive notions which do not contribute to a computational

---

[2] Thus, just as the core value of a proplet serves the double function of (i) representing the lexical semantics and (ii) determining the location for storage and retrieval, its continuation values serve the double function of (i) representing the compositional semantics and (ii) establishing a railroad system for the focus navigation, providing each proplet with only a limited choice of successor proplets.

reconstruction of intention in a cognitive agent. It plainly follows that "meant something" in the definiendum is unsuitable for computation as well.[3]

DBS treats the notion of intention computationally as the system-inherent (innate) drive to maintain the agent in a state of balance.[4] The principle of balance, in turn, is realized by means of *inferences*.

An inference consists of (i) an antecedent schema, (ii) a connective, and (iii) a consequent schema. Examples of connectives are **cm** (countermeasure), **impl** (implies), or **exec** (execute). An inference is triggered by a content matching its antecedent. Matching and binding the antecedent variables enable the consequent to derive an appropriately adapted blueprint for action (5.2.3).

The agent's balance may be disturbed by a nonlanguage perception, e.g., a sensation of hunger, or a language content, e.g., a reproach or a demand (hear mode, recognition), just as the countermeasure for regaining the agent's state of balance may be realized as a nonlanguage action, e.g., getting something to eat, or a language content, e.g., an apology or a request (speak mode, action).[5]

Because all behavior, including language behavior, is managed by the agent's autonomous control, it is essential for language *production* in communication:

5.1.2 THE FOURTH MECHANISM OF COMMUNICATION (MoC-4)

> The language as well as the nonlanguage behavior of a cognitive agent is driven by the goal of autonomous control to maintain a continuous state of balance vis à vis constantly changing external and internal environments. The success of autonomous control, short-, mid-, and long-term, is defined in terms of survival in the agent's ecological niche.

The niche includes the agent's social group. Within this group, maintaining a state of balance via language production ranges from ritualized customs like greeting with small talk to an exchange of opinion to a cry for help.

The overall goal of autonomous control to maintain the agent in a state of balance is a general method with many applications. For example, the method can be used not only to try to ensure the agent's survival in its ecological niche, but also for controlling the artificial agent's behavior for purposes of the user.

The artificial behavior may be reprogrammed by changing the "desired" values (defaults, *Sollwerte*) in the parameters of the agent's appraisal software.

---

[3] For further discussion of Grice's notion of meaning, see FoCL'99, Chap. 4, Example II.

[4] The cooperative behavior of social animals, e.g., ants in a colony, may also be described in terms of balance. Such a decentralized approach to behavior is in line with Brooks (1985).

[5] See Chap. 10 for an example which includes the use of balance to drive a dialogue.

This may be done highly selectively, at various levels of abstraction and in various domains. The resetting may also be done centrally by changing the cognition's center of gravity, thereby changing the agent's individual preferences and dislikes in one fell swoop, from nice to not-so-nice, from squeamish to matter-of-fact, from generous to stingy, etc., whatever has been implemented.

5.2 DBS Inferences for Maintaining Balance

Inferences may generally be described as deduction rules which derive new content from given content in a meaningful way. This holds for the deductive inferences in Symbolic Logic, which derive true conclusions (new content) from premises (given content). It also holds for DBS.

To maintain balance, DBS inferences take current and stored content, including the appraisal of previous action outcomes, as input, and compute blueprints for new action as output. There are three kinds, called R(eactor), D(eductor), and E(ffector) inferences.[6]

The condition 5.2.4 that the consequent of inference n must equal the antecedent of inference n+1 allows easy self-organization and highly adaptive data processing. The absence of rule names and rule packages facilitates definition and automation.

For readability, the following example of an inference chain is simplified as follows: (i) English words are used to represent proplets, (ii) coreference-by-address notation is omitted, and (iii) easily programmed details regarding the iteration of values in the variable restriction (line 3) are not included. The consequent of inference n always equals the antecedent of inference n+1.

5.2.1 CHAINING R, D, AND E INFERENCES

1. R:	β be hungry	K	cm	β eat food	K+1	
2. D:	β eat food	K+1	pre	β get food	K+2	
3. D:	β get food	K+2	down	β get α,	K+3	where $\alpha \in$ {apple, pear, salad, steak}
4. E:	β get α	K+3	exec	β locate α at γ	K+4	
5. E:	β locate α at γ	K+4	exec	β take α	K+5	
6. E:	β take α	K+5	exec	β eat α	K+6	
7. D:	β eat α	K+6	up	β eat food	K+7	

Each line begins with the step number, e.g., 1, followed by the kind of inference, e.g., R(eactor), the antecedent, e.g., β be hungry, the prn value of the antecedent, e.g., K, the connectives, e.g., cm, the consequent, e.g., β eat food,

---

[6] The terminology is intended to distinguish DBS inferences from the inferences of Symbolic Logic. For example, while a deductive inference like modus ponens is based on form, the reactor, deductor, and effector inferences of DBS may take content, domain, level of abstraction, etc., into account.

and the **prn** value of the consequent, e.g., K+1. Within a chain, any reoccurrence of a variable must be bound to a value at the Word Bank level equivalent to that of the initial occurrence of the variable. The equivalent value is either an address (5.2.3) or an indexical (5.2.5).

Step 1 is an R inference (defined in 5.2.3) with the connective **cm** (for countermeasure) and triggered by a sensation of hunger. Step 2 is a D inference with the connective **pre** (for precondition), while step 3 is a D inference for downward traversal (defined in 6.4.1) with the connective **down**. Steps 4, 5, and 6 are E inferences with the connective **exec** (for execute).

Step 4 may be tried iteratively for the instantiations of food provided by the consequent of step 3 (see the restriction on the variable α). If the agent cannot locate an apple, for example, it tries next to locate a pear, and so on. Individual food preferences of the agent may be expressed by the order of the elements in the variable restriction.

Step 7 is based on a D inference for upward traversal, defined in 6.4.4, with the connective **up**. This step is called the *completor* of the chain because the consequent of the chain-final inference, called the *completor consequent*, equals the consequent of step 1. The completor indicates the successful blueprint of a countermeasure to the imbalance characterized by the antecedent of the chain-initial reactor inference.

While R inferences are activated by triggers provided by the agent's recognition, external (e.g., *hot*) or internal (e.g., *hungry*), D and E inferences are usually initiated by other inferences which are already active. D(eductor) inferences establish meaning relations (Sect. 5.3), and are illustrated by synonymy (5.3.1), antonymy (5.3.2), cause and effect (5.3.3), summarizing (5.3.5), downward traversal (6.5.9), and upward traversal (6.5.12). One D inference may activate another D inference or an E inference.

E(ffector) inferences provide blueprints for the agent's action components.[7] Because E inferences connect central cognition with peripheral cognition, their definition has to be hand in glove with the robotic hardware they are intended to use.

A limiting case of a chain is a single R/E inference, such as the following:

5.2.2 ONE-STEP CHAIN BASED ON AN R/E INFERENCE

R/E: α feel full K **cm/exec** α stop eating K+1

Here the response to a deviation from balance results in a countermeasure which can be executed directly (i.e., without intervening D inferences).

---

[7] In robotics, effectors range from legs and wheels to arms and fingers.

DBS inferences resemble LA-think rules like 3.4.2. However, instead of merely navigating from one proplet to the next, an inference matches its antecedent to a content in the Word Bank in order to derive new content by means of its consequent. For example, using the format of pattern proplets and content proplets, the formal definition of the chain-initial R inference of 5.2.1 and its application to a content is as follows:

5.2.3 FORMAL DEFINITION AND APPLICATION OF A DBS INFERENCE

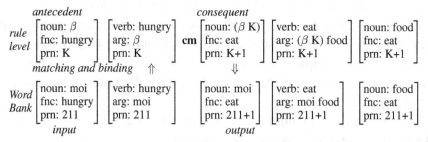

The inference is activated by the content *I am hungry* (input), which matches the antecedent. Utilizing the values to which the variables β and K are bound (i.e., moi and 211, respectively), the new content *I eat food* is derived by the consequent (output).

The inclusion of the antecedent's subject in the consequent by means of the address value (β K) excludes cases in which one agent is hungry and another one eats food – which would fail as an effective countermeasure. Because repeated reference by means of an indexical is exempt from the coreference-by-address method, the value corresponding to (β K) in the content derived by the consequent is simply moi rather than (moi 211).[8]

The consequent of a DBS inference may contain variables which do not appear already in the antecedent. An example is γ in the consequent of step 4 in 5.2.1. This is permissable as long as the co-domains of such variables are restricted to a certain range of possible values, as illustrated in 5.3.5.

The application of DBS inferences is governed by the following principle:

5.2.4 SEQUENTIAL INFERENCING PRINCIPLE (SIP)

Any two inferences x and y may be applied in sequence if, and only if, the consequent of x equals the antecedent of y.

This basic principle provides for a kind of self-organization (Kohonen 1988). Via equality of the antecedent or the consequent with other inferences, any

---
[8] For a more detailed discussion of indexicals, see Chaps. 10 and 11.

newly added inference is automatically connected. These bidirectional connections between inferences may be implemented efficiently by pointers.

Which inferences are activated at any given moment is determined by the agent's current state: recognition and inferencing produce a continuous stream of proplets, which are screened and used to select corresponding content in the Word Bank (subactivation, Sect. 5.4). This may apply at any level of abstraction and to any choice of domain. The extent of the subactivation is determined by such parameters as the available computing power, and the agent's personal interests and preferences.

High abstraction level patterns are used to select and subactivate (Sect. 5.4) relevant areas of content (domains). Intersection results in a substantially reduced set of proplets (search space reduction). Nevertheless, the result may still be a large number of inferences, each complete with corresponding contents and a next step for action. The triple consisting of (i) all contents triggering the antecedent, (ii) the inference, and (iii) the action blueprint derived by the consequent is called a **tia** triple (for trigger, inference, action).

The set of **tia** triples is evaluated in a cost-benefit manner, which results in further data reduction, i.e., a manageable number of blueprints for action. Which action is selected to be realized may often be a matter of random choice, especially when time is of essence. But it may also be the choice of long-term planning. Technically, the final step is a switch from the subjunctive mood to the indicative mood (Sect. 5.6), resulting in external action (realization).

The following example shows an inference chain (rule level) interacting with the Word Bank (content level). In line with SIP, the consequent of proposition n serves as the antecedent of proposition n+1.

5.2.5 New Content Derived by the Inference Chain 5.2.1

rule level: β be hungry K   **cm** β eat food K+1   **pre**
Word Bank: moi be hungry 211   moi eat food 212

      β get (food K+1)   K+2   **down** β get α K+3   **exec**
      moi get (food 212)   213   moi get apple   214

      β locate α at γ K+4   **exec** β take α K+5   **exec**
      moi locate (apple 214) at cupboard   215   moi take (apple 214)   216

      β (eat K+1) α K+6   **up** β (eat K+1) (food K+1) K+7
      moi (eat 212) (apple 214) 217   moi (eat 212) (food 212)   218

The four double lines should be read as one, i.e., as

rule level: **p1 cm   p2 pre   p3 down   p4 exec   p5 exec   p6 exec   p7 up   p8**
Word Bank: **q1   q2   q3   q4   q5   q6   q7   q8**.

The semiformal notation represents propositions by the core values of their proplets, e.g., **hungry**, and their **prn** value, e.g., **211**. Repeated core values referring to the same item are shown as addresses, e.g., (**food 212**), ensuring coreference between contents throughout the chain.

The chain is activated by a trigger situation, represented by the content **moi be hungry 211**, which matches the antecedent of the chain-initial reactor inference (shown in more detail in 5.2.3). The new content **moi eat food 212**, derived by the consequent, serves as the trigger content for the antecedent of the second inference (cf. step 2 in 5.2.1), and similarly for the remainder of the chain. The E inferences with the connective **exec** derive a sequence of blueprints for action, intended as a countermeasure.

The proplets derived by the completor consequent **211+7**, i.e., **moi (eat 212) (food 212) 218**, are added to the Word Bank at the current end of the respective token lines (i.e., the *now front*, cf. 4.1.1). Despite their **prn** value **218**, they are equivalent to the proplets derived earlier by the consequent of the initial R inference, i.e., **moi eat food 212**, due to their definition as the indexical **moi** and the pointer proplets (**eat 212**) and (**food 212**).

5.3 DBS Inferences for Meaning and Event Relations

The derivation of new content by means of DBS inferences may be used to formalize traditional relations of meaning. Standard examples from lexicography are *synonymy, antonymy, hypernymy, hyponymy, meronymy,* and *holonymy.* Consider the following example of a D inference implementing a synonymy:

5.3.1 INFERENCE RULE IMPLEMENTING A SYNONYMY

$$
\begin{bmatrix} \text{noun: abstract} \\ \text{fnc: } \alpha \\ \text{prn: K} \end{bmatrix} \textbf{ impl } \begin{bmatrix} \text{noun: summary} \\ \text{fnc: } \alpha \\ \text{prn: K+M} \end{bmatrix} \text{ where } \alpha \in \{\text{write, read, discuss, ...}\}
$$

According to this rule, **John wrote an abstract**, for example, implies that **John wrote a summary**. The restriction on the variable α specifies likely verbs, obtained from a corpus.

Next consider an example of a D inference implementing an antonymy:

5.3.2 INFERENCE RULE IMPLEMENTING AN ANTONYMY

$$
\begin{bmatrix} \text{adj: good} \\ \text{mdd: } \alpha \\ \text{prn: K} \end{bmatrix} \textbf{ impl } \begin{bmatrix} \text{adj: not bad} \\ \text{mdd: } \alpha \\ \text{prn: K+M} \end{bmatrix}
$$

When this inference is applied to an input content like **John had a good meal**, the result is the new content **John had a meal which was not bad** (cf. 6.5.9–6.5.11 for an analogous formal derivation).[9] Thus, the lexical meaning relations defined in 5.3.1–5.3.5 are formulated once as inferences at the rule level, but the results of their application may be written many times as new content to the *now front* of the Word Bank.

The meaning relations are coded solely in terms of core values connected by the usual semantic relations of structure, i.e., functor-argument and coordination. The new content coded by these relations may serve to trigger further inferences, for example, for the interpretation of nonliteral uses.[10]

Parallel to the meaning relations of lexicography are the event relations of philosophy, most notably *cause and effect*. These events exist independently of language, but due to the formal similarity between language content and context content in DBS (4.3.3), event relations may be formalized in the same way as the traditional meaning relations of lexicography:

5.3.3 INFERENCE RULE IMPLEMENTING A CAUSE AND EFFECT RELATION

$$
\begin{bmatrix} \text{noun: car} \\ \text{fnc: have} \\ \text{prn: K} \end{bmatrix}
\begin{bmatrix} \text{verb: have} \\ \text{arg: car no fuel} \\ \text{prn: K} \end{bmatrix}
\begin{bmatrix} \text{noun: no fuel} \\ \text{fnc: have} \\ \text{prn: K} \end{bmatrix}
\ \mathbf{impl}\
\begin{bmatrix} \text{noun: (car K)} \\ \text{fnc: no start} \\ \text{prn: K+M} \end{bmatrix}
\begin{bmatrix} \text{verb: no start} \\ \text{arg: (car K)} \\ \text{prn: K+M} \end{bmatrix}
$$

This DBS inference formalizes the relation **If a car has no fuel then it does not start**, where fuel may be instantiated by gasoline, electricity, methane, etc.

The descriptive power of the DBS inferencing method may be shown further by condensing a complex content into a meaningful summary. As an example, consider the following short text, formally derived in Chaps. 13 (hear mode) and 14 (speak mode) of NLC'06:

> The heavy old car hit a beautiful tree. The car had been speeding. A farmer gave the driver a lift.

After syntactic-semantic parsing by LA-hear, the content of this modest text is stored in the agent's Word Bank. A reasonable summary of the content would be *car accident*. This summary may be represented in the agent's Word Bank as follows:

---

[9] The rules 5.3.1 and 5.3.2 illustrate how traditional meaning relations may be handled in DBS. Whether or not the inverse relations hold as well is an empirical question which may be answered by distinguishing between unidirectional and bidirectional inferences. Given that the inferences in question are part of an individual agent's cognition, they do not have the status of universal truths. Instead, they are evaluated and adapted in terms of their utility for the agent's day-to-day survival.

[10] Thus, the hypernymy inference **dog inst**(antiates) **animal** may derive the new content **Fido is an animal** from **Fido is a dog**. In the hear mode, this new content may be used to infer from the language expression **The animal is tired** that the intended utterance meaning is **Fido is tired** (cf. FoCL, Sect. 4.5; NLC'06, Sect. 5.4) – and conversely in the speak mode.

5.3.4 RELATING SUMMARY car accident TO TEXT

	member proplets	owner proplets

...

$$\begin{bmatrix} \text{noun: accident} \\ \text{mdr: (car 1)} \\ \text{prn: 67} \end{bmatrix} \quad \dots [\text{core: accident}]$$

...

$$\dots \begin{bmatrix} \text{noun: car} \\ \text{fnc: hit} \\ \text{prn: 1} \end{bmatrix} \begin{bmatrix} \text{noun: (car 1)} \\ \text{fnc: speed} \\ \text{prn: 2} \end{bmatrix} \dots \qquad \begin{bmatrix} \text{noun: (car 1)} \\ \text{mdd: accident} \\ \text{prn: 67} \end{bmatrix} \quad \dots [\text{core: car}]$$

...

$$\dots \begin{bmatrix} \text{verb: hit} \\ \text{arg: car tree} \\ \text{prn: 1} \end{bmatrix} \dots \qquad \qquad \qquad \dots [\text{core: hit}]$$

$$\dots \begin{bmatrix} \text{verb: speed} \\ \text{arg: (car 1)} \\ \text{prn: 2} \end{bmatrix} \qquad \qquad \dots [\text{core: speed}]$$

...

The connection between propositions 1 and 2 and summary 67 is coreference-by-address. It is based on the original car value in proposition 1, the corresponding address value (car 1) in proposition 2, and similarly in summary 67.

The summary consists of the *accident* proplet and a *car* pointer proplet. They share the prn value 67, and are connected by the modifier-modified relation. The summary and the original text are connected by the address value (car 1), which serves as the core value of the second *car* proplet as well as the mdr (modifier) value of the *accident* proplet.

How is the summary automatically derived from the text? The summary-creating inference deriving the new content with the prn value 67 is formally defined as the following D(eductor) inference rule, shown with the sample input and output of 5.3.4 at the Word Bank (content) level:

5.3.5 SUMMARY-CREATING D INFERENCE

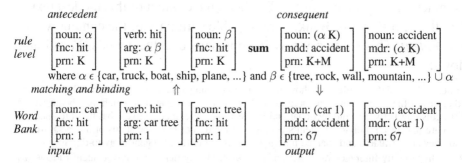

Antecedent and consequent are related by the connective **sum**(marize).

In the rule, the possible values which α and β may be bound to during matching are restricted by the co-domains of these variables: the restricted variable α generalizes the summary-creating inference to different kinds of accidents, e.g., *car accident, truck accident,* etc., while the restricted variable β limits the objects to be hit to trees, rocks, etc., as well as cars, trucks, etc. Any content represented by the proplet *hit* with a subject and an object proplet satisfying the variable restrictions of α and β, respectively, will be automatically (i) summarized as an accident of a certain kind where (ii) the summary is related to the summarized by means of an address value, here (car 1), thus fulfilling the condition that the data in a content-addressable memory not be modified.

By summarizing content into shorter and shorter versions, there emerges a hierarchy providing retrieval relations for upward or downward traversal (Sect. 6.5). Upward traversal supplies the agent with more general notions at a higher level of abstraction, while downward traversal supplies more concrete instantiations at a lower level of abstraction. Either kind may be used to access and to apply inferences defined at another level of abstraction, and to subsequently return to the original level.

5.4 Subactivation and Intersection

The amount of data in the agent's memory may be very large. For efficiently finding contents in memory which match the agent's current situation DBS introduces a database mechanism called *subactivation*. It is a kind of guided association, which continuously accompanies the agent's current cognition with corresponding content stored in the Word Bank.

Subactivation works like a dragnet, pulled by the concepts activated by the agent's current recognition and inferencing, and providing them with relevant experiences and knowledge from the agent's past. As a form of association,[11] subactivation results in a mild form of selective attention.

Intuitively, subactivation may be viewed as highlighting an area of content at half-strength, setting it off against the rest of the Word Bank, but such that exceptional evaluations are still visible as brighter spots. In this way, the agent will be alerted to potential threats or opportunities even in current situations which would otherwise seem innocuous.

An elementary subactivation consists of three steps and may be of a a primary and a secondary degree. The first step is subactivating the token line

---

[11] Such as associating a certain place with a happy or an unhappy memory.

which corresponds to a trigger concept provided by the agent's current situation. Consider the following example:

5.4.1 TRIGGER CONCEPT SUBACTIVATING CORRESPONDING TOKEN LINE

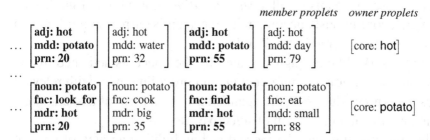

The trigger concept *hot* provided by the agent's external recognition matches the core value hot of an owner record in the agent's Word Bank. This subactivates the associated token line.

However, if a continuous sequence of trigger concepts were to always subactivate complete token lines, the resulting amount of data would be too large to be useful. Therefore, the second step is to use the semantic relations of functor-argument and coordination connecting the incoming concepts for restricting subactivation via *intersection* of the relevant token lines. In this way, the more semantically connected the concepts coming in, the more narrow and specific the subactivated data (search space reduction).

For example, if the agent's durrent recognition relates *hot* and *potato* as modifier-modified, the token lines of *hot* and *potato* might contain the following intersections, indicated typographically by bold face:

5.4.2 INTERSECTING TOKEN LINES FOR *hot* AND *potato*

	member proplets			owner proplets
...	$\begin{bmatrix} \textbf{adj: hot} \\ \textbf{mdd: potato} \\ \textbf{prn: 20} \end{bmatrix}$ $\begin{bmatrix} \text{adj: hot} \\ \text{mdd: water} \\ \text{prn: 32} \end{bmatrix}$	$\begin{bmatrix} \textbf{adj: hot} \\ \textbf{mdd: potato} \\ \textbf{prn: 55} \end{bmatrix}$ $\begin{bmatrix} \text{adj: hot} \\ \text{mdd: day} \\ \text{prn: 79} \end{bmatrix}$		[core: hot]
...	$\begin{bmatrix} \textbf{noun: potato} \\ \textbf{fnc: look\_for} \\ \textbf{mdr: hot} \\ \textbf{prn: 20} \end{bmatrix}$ $\begin{bmatrix} \text{noun: potato} \\ \text{fnc: cook} \\ \text{mdr: big} \\ \text{prn: 35} \end{bmatrix}$	$\begin{bmatrix} \textbf{noun: potato} \\ \textbf{fnc: find} \\ \textbf{mdr: hot} \\ \textbf{prn: 55} \end{bmatrix}$ $\begin{bmatrix} \text{noun: potato} \\ \text{fnc: eat} \\ \text{mdd: small} \\ \text{prn: 88} \end{bmatrix}$		[core: potato]

The example contains two intersections, each consisting of two proplets sharing (i) a prn value and (ii) the modifier-modified relation between *hot* and *potato*. The intersections differ from each other in their respective prn values, 20 and 55, and the fnc values of the nouns, look_for and find.

The third step of a primary subactivation consists of completing an intersection into a full proposition. It is based on subactivating the intrapropositional semantic relations (spreading activation, Quillian 1968). For example, using the functor-argument coded by the leftmost proplets in 5.4.2, the intersection may be completed into a proposition: .

5.4.3 COMPLETING AN INTERSECTION BY SPREADING ACTIVATION

$$
\begin{bmatrix} \text{noun: John} \\ \text{fnc: look\_for} \\ \text{prn: 20} \end{bmatrix}
\begin{bmatrix} \text{verb: look\_for} \\ \text{arg: John, potato} \\ \text{pc: cook 19} \\ \text{nc: eat 21} \\ \text{prn: 20} \end{bmatrix}
\begin{bmatrix} \textbf{noun: potato} \\ \textbf{fnc: look\_for} \\ \textbf{mdr: hot} \\ \textbf{prn: 20} \end{bmatrix}
\begin{bmatrix} \textbf{adj: hot} \\ \textbf{mdd: potato} \\ \textbf{prn: 20} \end{bmatrix}
$$

This primary subactivation is based on the addresses potato 20 in the *hot* proplet, look_for 20 in the *potato* proplet, and John 20 in the *look_for* proplet.

While a primary subactivation utilizes the intrapropositional relations of functor-argument and coordination (cf. NLC'06, Chaps. 6 and 8), a secondary subactivation is based on the corresponding extrapropositional relations (cf. NLC'06, Chaps. 7 and 9). For example, using the pc (previous conjunct) and nc (next conjunct) values of the *look_for* proplet in 5.4.3, a secondary subactivation may spread from John looked for a hot potato to the predecessor and successor propositions with the verb values cook and eat and the prn values 19 and 21, respectively.[12]

The degree of subactivation that is automatically selected at any current moment, including no subactivation at all, depends on the computational resources available. These depend on the agent's interests and current preoccupation. Compare, for example, running for one's life and a leisurely walk, both in the same park: the very same triggers will be completely ignored in the first case, but may result in rich subactivations in the second.

5.5 Analogical Models for Problem Solving

Sect. 5.2 showed a *data-driven*[13] application of the inference chain 5.2.1 (forward chaining). Let us consider now a *goal-driven*[14] application of this same inference chain (backward chaining).[15] While a data-driven application is for instinctive, habitual, and rote behavior, a goal-driven application is for nonroutine behavior such as analogical reasoning and problem solving. The backward-chaining of a goal-driven inference application makes heavy use of subactivation and intersection.

[12] Thus, the movie title All the President's Men (Pakula 1976) will likely activate Couldn't put Humpty Dumpty Together Again as a continuation, referring to R. M. Nixon. In fiction, our notion of triggering a spreading subactivation is illustrated by the madeleine experience of Proust (1913), which brings back an almost forgotten area of what he calls "l'édifice immense du souvenir."

[13] So-called because the inference chain is triggered by data matching the initial antecedent.

[14] So-called because the inference chain is triggered by a goal matching a consequent. A goal-driven application is a form of case-based reasoning (Schank 1982). See also Muñoz-Avila et al. (2010).

[15] We follow the standard terminology in computer science, even though from the viewpoint of DBS the input to goal-driven inferencing, i.e., the goal, is data too – no different from that triggering data-driven inferencing.

For example, the observation of another agent eating an apple matches the consequent **K+6** of the penultimate inference in 5.2.1. This observation may set in motion a search, to determine where the apple came from in order get one for oneself as well.

The search starts with intersections between (i) the agent's current task and (ii) potential countermeasures stored in the agent's memory, observed or self-performed. Of the available countermeasures, the one (a) best matching the agent's current situation and (b) with the best outcome is automatically selected. If no such countermeasure is available, the agent's options are some additional inferencing and ultimately trial and error.

If an appropriate countermeasure has been found, the second step is a *transfer* of the content in question. The transfer replaces the agent of the remembered countermeasure by **moi**, and provides new **prn** and address values. The result is a new content, written to the *now front* and serving as a blueprint for an action sequence suitable for reestablishing the agent's balance.

Assume, for example, that the agent is alone in Mary's house – which serves as a trigger (5.4.1) subactivating the token line of *Mary* in the agent's Word Bank. Furthermore, the agent is hungry, which triggers the *hungry-eat* inference chain 5.2.1. The constant *eat* in the consequent of the completor inference of the chain subactivates the corresponding token line, resulting in intersections between the *Mary* and the *eat* token lines such as the following:

5.5.1 TWO *Mary eat* INTERSECTIONS

$$
\begin{bmatrix} \text{noun: (Mary 25)} \\ \text{fnc: eat} \\ \text{prn: 48} \end{bmatrix}
\begin{bmatrix} \text{verb: eat} \\ \text{arg: (Mary 25) (\textbf{apple 46})} \\ \text{pc: take 47} \\ \text{prn: 48} \end{bmatrix}
\qquad
\begin{bmatrix} \text{noun: (Mary 25)} \\ \text{fnc: eat} \\ \text{prn: 82} \end{bmatrix}
\begin{bmatrix} \text{verb: eat} \\ \text{arg: (Mary 25) (müsli 80)} \\ \text{pc: prepare 81} \\ \text{prn: 82} \end{bmatrix}
$$

In other words, the agent remembers Mary in her house once eating an apple and once eating müsli.

The two proplets in each intersection share a **prn** value, namely 48 and 82, respectively, and are in a semantic relation of structure, namely functor-argument. In both intersections, the verb proplet *eat* provides at least one yet unrealized intrapropositional continuation, namely **(apple 46) 48** in the first and **(müsli 80) 82** in the second. Following the continuation in the first intersection results in the following primary (5.4.3) subactivation:

5.5.2 SUBACTIVATION SPREADING FROM *Mary eat* TO *Mary eat apple*

$$
\begin{bmatrix} \text{noun: (Mary 25)} \\ \text{fnc: eat} \\ \text{prn: 48} \end{bmatrix}
\begin{bmatrix} \text{verb: eat} \\ \text{arg: (Mary 25) (\textbf{apple 46})} \\ \text{pc: \textbf{take 47}} \\ \text{prn: 48} \end{bmatrix}
\begin{bmatrix} \text{noun: (apple 46)} \\ \text{fnc: eat} \\ \text{\textbf{eval: attract}} \\ \text{prn: 48} \end{bmatrix}
$$

The *(apple 46) 48* proplet contains the feature [eval: attract]. Assuming that the corresponding subactivation for the second intersection in 5.5.1 happens to evaluate the *(müsli 80) 82* proplet as [eval: avoid][16] (not shown), the agent will use only the first, and not the second, subactivation as a trigger situation to activate a consequent in the inference chain 5.2.1:

5.5.3 STORED CONTENT MATCHING CONSEQUENT IN INFERENCE CHAIN

| *rule level:* | β be hungry **K cm** | β eat food K+1 | **pre** |
| *Word Bank* | # | # | |

| | β get food K+2 **down** | β get α K+3 | **exec** |
| | # | # | |

| | β locate α at γ K+4 **exec** | β take α K+5 | **exec** |
| | # | # | |

| | β eat α K+6 **up** | β eat food K+7 | |
| | (Mary 25) eat (apple 46) 48 | # | |

The trigger content **(Mary 25) eat (apple 46) 48** matches the consequent **K+6** of the penultimate inference. All other inference parts in the chain do not have matching contents at this point, indicated by #.

Next, the matching content **(Mary 25) eat (apple 46) 48** is used for a secondary (i.e., extrapropositional) subactivation. It includes propositions which precede at the agent's content level, and which contain nouns coreferent with **(Mary 25)** and with **(apple 46)**. Pattern matching selects subactivated propositions which fit antecedents and consequents preceding β **eat** α **K+6** in the inference chain 5.2.1, resulting, for example, in the following correlation of inferences and contents:

5.5.4 EXTENDING MATCHING CONTENT BY SECONDARY SUBACTIVATION

| *rule level:* | β be hungry **K cm** | β eat food K+1 | **pre** |
| *Word Bank:* | # | # | |

| | β get food K+2 **down** | β get α K+3 | **exec** |
| | # | # | |

| | β locate α at γ | K+4 **exec** | |
| | (Mary 25) locate apple at cupboard | 46 | |

| | β take α | K+5 **exec** | |
| | (Mary 25) take (apple 46) | 47 | |

| | β eat α K+6 **up** | β eat food K+7 | |
| | (Mary 25) eat (apple 46)   48 | # | |

---

[16] The assumed evaluations reflect the agent's preference for eating apples over eating müsli.

At this point, the chain with partially[17] matching contents may be used by the hungry agent as a model for regaining balance. All that is required is to replace Mary by moi (transfer) and to derive new content with new prn and address values as blueprints for action:

5.5.5 TRANSFER AND COMPLETION

| *rule level:* | β be hungry | **K cm** | β eat food | K+1 | **pre** |

Word Bank: #   #

| β get food | K+2 **down** | β get α | K+3 | **exec** |

         #   #

| β | locate | α | at γ | K+4 | **exec** |
| moi | locate apple | at cupboard | 91 |

| β | take | α | | K+5 | **exec** |
| moi | take | (apple 91) | | 92 |

| β | eat | α | K+6 **up** | β eat food | K+7 |
| moi eat | (apple 91) | 93 | moi eat food | 94 |

The content propositions following the **exec** connective at the rule level, i.e., 91–93, are blueprints for action.[18] They are ready for realization, but the final decision about whether or not they are actually passed on to the agent's action components (cf. diagram 4.5.3, interface 6) is still open.

5.6 Subjunctive Transfer and Managing the Data Stream

The agent's decision of whether or not to realize a current blueprint for action is implemented as a change of what is called the *verbal mood* in linguistics. In English, the verbal moods are the indicative, the subjunctive, and the imperative. In DBS, they are formally implemented as the ind, sbjv, and impv values of the sem(antics) attribute in verb proplets (3.5.6).

Content derived by the agent's recognition is indicative. A blueprint derived by inferencing, but not yet passed to the agent's action components, is subjunctive. Old content may be of any verbal mood and does not initiate any action directly. However, activated old content may initiate the derivation of new content by matching the initial antecedent (forward chaining) or a consequent (backward chaining) of an inference chain. The propositions of this

---

[17] If the agent were to assume (unnecessarily) that Mary must have been hungry (thus supplying content all the way to the initial R inference), then this would correspond to an abductive inference in logic. The point is that observing Mary locating, taking, and eating an apple is sufficient for the purpose at hand.

[18] If the propositions 91–93 have been executed successfully, proposition 94 completes the counter-measure for regaining balance.

new content may also be of any verbal mood, but only imperative blueprints following the **exec** connective are passed to the agent's action components for immediate realization.

A subjunctive blueprint may be revised into an imperative blueprint by means of the following inference, which (i) copies the content using addresses and new **prn** values, (ii) changes the verbal mood value from subjunctive to imperative, and (iii) uses the connective **mc**, for *mood change*:

5.6.1 INFERENCE CHANGING SUBJUNCTIVE TO IMPERATIVE CONTENT

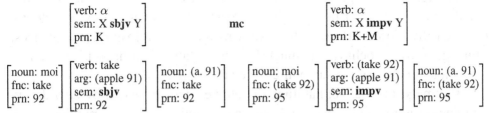

The content matching the antecedent is identical to proposition **92** in 5.5.5, except that it is represented explicitly by proplets. Whether or not the **mc** inference is applied and the action sequence is actually attempted depends on agent-internal and -external circumstances. An internal circumstance leading to the application would be a high need for a countermeasure, while a low need may leave the application unexecuted. An external circumstance stopping the application may be an override, such as a ringing phone.

If the **mc** inference is applied, the success of the attempted action sequence will depend on whether the agent's environment provides the necessary preconditions. For example, the countermeasure derived in 5.5.5 will be successful only if proposition **91** turns out to hold in the agent's current situation.

In addition to the switching of a subjunctive blueprint into an imperative one, there is also the switching of an indicative content into the subjunctive, especially in combination with transfer. An example of such a *subjunctive transfer* is *empathy*, i.e., an agent's ability to share another being's emotions and feelings. For example, when agent A observes in a movie how actor B is being attacked by a monster, A may empathize with B by subjunctive transfer, i.e., by tentatively replacing B with A in the content transported by the movie (cf. transition from Mary in 5.5.4 to moi in 5.5.5). In this way, A may experience the agony of B while knowing that there is no danger – experienced as a pleasant thrill.[19] This is easily modeled in an artificial agent.

---

[19] A related phenomenon is the identification with the fortunes of a favorite sports club.

Subjunctive transfer is also essential for constructing the agent's *discourse model* in dialogue (cf. Dominey and Warneken, in press). Consider, for example, two agents discussing going on a vacation together. Their respective discourse models include not only knowledge of the subject matter, e.g., where to go, how to get there, where to stay, what to do there, how to pay for it, etc., but also knowledge of the other agent. While knowledge of the subject matter is subactivated by means of intersection, knowledge of the other agent's cognitive state is based also on subjunctive transfer.

Even though subjunctive transfer will provide maximum information about another agent's viewpoints, feelings, tolerances, preferences, dislikes, demands, etc., it may not be enough. For example, agent A may inadvertently hit a sensitive spot of B, hitherto unknown and with explosive consequences. Thus, goal-driven behavior is much less predictable than data-driven behavior – though both use the same inference chain, e.g., 5.2.1.

In summary, instinctive, habitual, or rote behavior provided by data-driven inferencing is the same over and over again and is therefore predictable in another agent. Another agent's goal-driven behavior,[20] in contrast, is unpredictable because it depends on data stored in the other agent's memory and therefore not directly accessible to the partner in discourse.

The DBS inferencing described is continuously triggered by subactivation and intersection, which in turn is activated by current language and nonlanguage recognition. The result is a stream of new inference data written to the *now front* – in addition to current language and nonlanguage recognition. This raises the more general question of how to avoid overflow of the agents' large, but finite memory.

Given that a content-addressable memory cannot be changed, incoming data should be selected and cleaned up prior to storage. Following the natural prototype, this may be done by providing a short-term and a long-term memory, and by doing selection and cleanup in short-term memory prior to long-term storage. For example, subjunctive and imperative action sequences may be derived in short-term memory, but only the imperative sequences need to be remembered long-term.

The other method to stave off memory overflow, also following the natural prototype, is forgetting. Though a limiting case of data change (and therefore

---

[20] Goal-driven behavior is important to nouvelle AI in general (cf. Braitenberg 1984, Brooks 1991) and BEAM robotics (for Biology, Electronics, Aesthetics, Mechanics) in particular (cf. Tilden and Hasslacher 1996). However, while BEAM has the goal to model goal driven behavior with the simplest means possible, based on analogical sensors without microprocessors, DBS uses representations of content in a database. This is because DBS starts out from natural language, while BEAM robotics proceeds from insect behavior.

not really permitted in a content-addressable memory), forgetting is functionally acceptable if the items deleted do not result in any (or result in only a few) dead ends in the associative data network. As a solution, Anderson (1983) proposed a frequency-based approach, namely to bleach those data from memory which are never or very rarely activated by the cognitive operations of the agent.[21] The procedure resembles garbage collection in programming languages like LISP, Java, or Oberon.

In a Word Bank, the many dispersed little spaces created by bleaching may be made reusable by pushing the proplets in a token line to the left, like pearls on a string (defragmentation). In this way, the intrinsic ordering of the proplets' storage positions is maintained, and the contiguous spaces reclaimed on the righthand end of the token lines are suitable for new storage at the *now front*. Furthermore, the declarative addresses coding inter-proplet relations (Sect. 4.4) provide a straightforward solution to the recomputation of the pointers to new physical storage locations, necessary after bleaching and defragmentation. The periods in which bleaching, defragmentation, and recomputation are performed in artificial agents may be likened to the periods of sleep in natural agents.

The handling of the data stream inspired by natural phenomona such as sleep and forgetting may be complemented by purely technical means. Just as an aircraft may be designed to maximize payload, range, speed, and profit with little or no resemblance to the natural prototypes (e.g., birds), there may be a software routine which is applied whenever the amount of memory in the Word Bank reaches a certain limit; the procedure compresses the currently oldest content segment (containing, for example, all proplets with prn values between 30 000 and 32 000) and moves it into secondary storage without any deletion (and thus without any need to choose).[22]

The only downside to data in secondary storage would be a slight slowdown in initial activation. Depending on the application, either an artificial agent's overall memory space may be sized from the beginning to accommodate the amount of data expected for an average lifetime (for example, in a robot on a space mission), or the necessary hardware may be expanded incrementally as needed.

---

[21] For example, we tend to forget minor errors like "I didn't put the garlic on the lamb!" However, the intensity of a memory in its historical context must also be taken into account in order to handle examples like Proust's madeleine episode.

[22] More differentiated routines are easily imaginable, depending on the application.

6. Mystery Number Five: Learning

The wide variety of DBS grammar rules and DBS inferences described so far would be of limited use if they had all to be defined and adjusted by hand. Therefore, the fifth mystery of natural communication is how to model (i) *adaptation* during evolution in phylogenesis and ontogenesis, including language acquisition, and (ii) *learning* as an improvement of the agent's survival skills in a changing environment.

6.1 Fixed Behavior Agents

The phylogenetic and ontogenetic evolutions of an agent's cognition begins with fixed behavior, based on the fixed action patterns (**FAPs**) of ethology (Campbell 1996). As an abstract example consider an agent which can perceive no more than three external stimuli, namely a red, a green, and a blue light (recognition), and can perform no more than three kinds of external motion, namely straight, left, and right (action). Furthermore, a red light triggers a straight, left, right, straight, right, left sequence, a green light triggers a straight, left, straight, left, straight, left sequence, and a blue light triggers a straight, right, straight, right, straight, right sequence. Without going into the angle, length, etc., of each kind of step, a graphical representation of these action sequences may be shown roughly as follows:

6.1.1 MOTION PATTERNS OF A FIXED BEHAVIOR AGENT

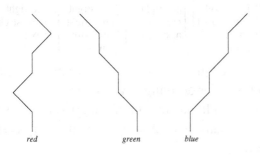

red   green   blue

Each of these fixed stimulus-response pairs may be utilized for (i) conspecific communication, e.g., as a mating dance or a greeting, or for (ii) non-communicative behavior, e.g., hiding by wiggling into the sand or as a pattern of flight.

In adaptive behavior (Sect. 5.5), recognition action events are recorded in the temporal order of their arrival in the Word Bank. This is indicated by using natural numbers as prn values, as in the extrapropositional coordination 3.2.5.

In fixed behavior, in contrast, recognition action sequences are used as patterns. When triggered, they are repeated without keeping track of individual instances.[1] This is indicated by using prn values consisting of two parts, a letter common to all proplets of the sequence and subscripted natural numbers specifying the order within the sequence. Consider the following set of proplets, which codes the first behavior pattern in 6.1.1:

6.1.2 CODING MOTION TRIGGERED BY red AS SET OF PROPLETS

$$
\begin{bmatrix} \text{rec: red} \\ \text{prev:} \\ \text{next: strght} \\ \text{prn: } x_1 \end{bmatrix}
\begin{bmatrix} \text{act: strght} \\ \text{prev: red} \\ \text{next: left} \\ \text{prn: } x_2 \end{bmatrix}
\begin{bmatrix} \text{act: left} \\ \text{prev: strght} \\ \text{next: right} \\ \text{prn: } x_3 \end{bmatrix}
\begin{bmatrix} \text{act: right} \\ \text{prev: left} \\ \text{next: strght} \\ \text{prn: } x_4 \end{bmatrix}
\begin{bmatrix} \text{act: strght} \\ \text{prev: right} \\ \text{next: right} \\ \text{prn: } x_5 \end{bmatrix}
\begin{bmatrix} \text{act: right} \\ \text{prev: strght} \\ \text{next: left} \\ \text{prn: } x_6 \end{bmatrix}
\begin{bmatrix} \text{act: left} \\ \text{prev: right} \\ \text{next:} \\ \text{prn: } x_7 \end{bmatrix}
$$

The sequence begins with a *red* proplet, which represents the stimulus activating the motion pattern. The next attribute of the *red* proplet has the value strght, the next attribute of the *strght* proplet has the value left, etc.

Similarly for the other two motion patterns of 6.1.1:

6.1.3 CODING MOTION TRIGGERED BY green AS SET OF PROPLETS

$$
\begin{bmatrix} \text{rec: green} \\ \text{prev:} \\ \text{next: strght} \\ \text{prn: } y_1 \end{bmatrix}
\begin{bmatrix} \text{act: strght} \\ \text{prev: green} \\ \text{next: left} \\ \text{prn: } y_2 \end{bmatrix}
\begin{bmatrix} \text{act: left} \\ \text{prev: strght} \\ \text{next: strght} \\ \text{prn: } y_3 \end{bmatrix}
\begin{bmatrix} \text{act: strght} \\ \text{prev: left} \\ \text{next: left} \\ \text{prn: } y_4 \end{bmatrix}
\begin{bmatrix} \text{act: left} \\ \text{prev: strght} \\ \text{next: strght} \\ \text{prn: } y_5 \end{bmatrix}
\begin{bmatrix} \text{act: strght} \\ \text{prev: left} \\ \text{next: left} \\ \text{prn: } y_6 \end{bmatrix}
\begin{bmatrix} \text{act: left} \\ \text{prev: strght} \\ \text{next:} \\ \text{prn: } y_7 \end{bmatrix}
$$

6.1.4 CODING MOTION TRIGGERED BY blue AS SET OF PROPLETS

$$
\begin{bmatrix} \text{rec: blue} \\ \text{prev:} \\ \text{next: strght} \\ \text{prn: } z_1 \end{bmatrix}
\begin{bmatrix} \text{act: strght} \\ \text{prev: blue} \\ \text{next: right} \\ \text{prn: } z_2 \end{bmatrix}
\begin{bmatrix} \text{act: right} \\ \text{prev: strght} \\ \text{next: strght} \\ \text{prn: } z_3 \end{bmatrix}
\begin{bmatrix} \text{act: strght} \\ \text{prev: right} \\ \text{next: right} \\ \text{prn: } z_4 \end{bmatrix}
\begin{bmatrix} \text{act: right} \\ \text{prev: strght} \\ \text{next: strght} \\ \text{prn: } z_5 \end{bmatrix}
\begin{bmatrix} \text{act: strght} \\ \text{prev: right} \\ \text{next: strght} \\ \text{prn: } z_6 \end{bmatrix}
\begin{bmatrix} \text{act: right} \\ \text{prev: strght} \\ \text{next:} \\ \text{prn: } z_7 \end{bmatrix}
$$

The storage and retrieval of these sequences in a Word Bank is content-addressable, but their activation is not for deriving new content.

Therefore, fixed behavior uses only a fixed amount of memory, in contradistinction to adaptive behavior. This means that a non-writable memory is sufficient for modeling fixed behavior.

---

[1] Webb et al. (2009) classify increasingly powerful software systems of behavior in Fig. 1.2.

The sequences are performed by the following LA-grammar, called LA-act1 (for action). The definition begins with the variables:

6.1.5 VARIABLE DEFINITION OF THE LA-ACT1 GRAMMAR

T_n ϵ {red, green, blue} and n ϵ {1, 2, 3, ...}
M1 ϵ {strght, left, right}
M2 ϵ {strght, left, right}
K ϵ {x_i, y_i, z_i, ...} and i ϵ {1, 2, 3, ...}

The variables are restricted to the values in the corresponding sets. The rule system of LA-act1 is defined as follows:

6.1.6 RULE SYSTEM OF THE LA-ACT1 GRAMMAR

$ST_S =_{def}$ { ($[\text{rec: } T_n]$ {Rule_0, Rule_1}) }

Rule_0   {Rule_0, Rule_1}

$\begin{bmatrix} \text{rec: } T_n \\ \text{next: } T_{n+1} \\ \text{prn: } K_i \end{bmatrix}$ $\begin{bmatrix} \text{rec: } T_{n+1} \\ \text{prev: } T_n \\ \text{prn: } K_{i+1} \end{bmatrix}$ output position nw

Rule_1   {Rule_2}

$\begin{bmatrix} \text{rec: } T_n \\ \text{next: } M1 \\ \text{prn: } K_i \end{bmatrix}$ $\begin{bmatrix} \text{act: } M1 \\ \text{prev: } T_n \\ \text{prn: } K_{i+1} \end{bmatrix}$ output position nw

Rule_2   {Rule_2}

$\begin{bmatrix} \text{act: } M1 \\ \text{next: } M2 \\ \text{prn: } K_i \end{bmatrix}$ $\begin{bmatrix} \text{act: } M2 \\ \text{prev: } M1 \\ \text{prn: } K_{i+1} \end{bmatrix}$ output position nw

$ST_F =_{def}$ {($[\text{next: }]$ rp_{Rule_2})}

The LA-act1 grammar automatically starts navigating through the agent's Word Bank whenever its start state ST_S is satisfied. This is the case whenever the agent's recognition provides one of the values of the trigger variable T_n (6.1.5), i.e., whenever the agent perceives a **red**, **green**, or **blue** light.

The rule package of the start state ST_S calls Rule_0 and Rule_1. Rule_0 may be applied recursively to parse complex stimuli, like **red green** or **red red red**. Rule_1 moves from the end of the stimulus to the beginning of the motion sequence and calls Rule_2. Rule_2 calls itself and completes the motion sequence. After that the fixed behavior agent comes to rest until it is triggered by another stimulus.

As an example of a rule application consider Rule_1, called from the start state (rather than from Rule_0) and applied to the input proplet *red*:

6.1.7 Applying Rule_1 of LA-ACT1 to a red trigger

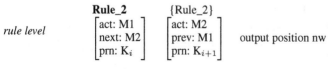

rule level

Rule_1 {Rule_2}

$$\begin{bmatrix} \text{rec: } T_n \\ \text{next: M1} \\ \text{prn: } K_i \end{bmatrix} \quad \begin{bmatrix} \text{act: M1} \\ \text{prev: } T_n \\ \text{prn: } K_{i+1} \end{bmatrix} \qquad \text{output position nw}$$

matching and binding

Word Bank level

$$\begin{bmatrix} \text{rec: red} \\ \text{prev:} \\ \text{next: strght} \\ \text{prn: } x_1 \end{bmatrix} \quad \begin{bmatrix} \text{act: strght} \\ \text{prev: red} \\ \text{next: left} \\ \text{prn: } x_2 \end{bmatrix}$$

By matching the first proplet pattern at the rule level to the corresponding first proplet at the Word Bank level, the trigger variable T_n is bound to **red**, the motion variable M1 is bound to **strght**, and the **prn** variable K_i is bound to x_1. In this way, the variables in the second pattern at the rule level are provided with values which are sufficient to retrieve (navigate to, activate) the second proplet shown at the Word Bank level,[2] i.e., the proplet with the core value **strght** and the **prn** value x_2.

After the successful application of Rule_1, its rule package calls Rule_2, which takes the second proplet of 6.1.7 (Word Bank level) as input and continues the navigation through the stimulus-response sequence in question:

6.1.8 Applying Rule_2 of LA-ACT1 to a strght motion

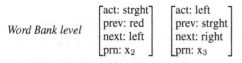

rule level

Rule_2 {Rule_2}

$$\begin{bmatrix} \text{act: M1} \\ \text{next: M2} \\ \text{prn: } K_i \end{bmatrix} \quad \begin{bmatrix} \text{act: M2} \\ \text{prev: M1} \\ \text{prn: } K_{i+1} \end{bmatrix} \qquad \text{output position nw}$$

matching and binding

Word Bank level

$$\begin{bmatrix} \text{act: strght} \\ \text{prev: red} \\ \text{next: left} \\ \text{prn: } x_2 \end{bmatrix} \quad \begin{bmatrix} \text{act: left} \\ \text{prev: strght} \\ \text{next: right} \\ \text{prn: } x_3 \end{bmatrix}$$

Again, the variables of the first proplet pattern at the rule level are bound to the corresponding values of the corresponding proplet at the Word Bank level (which is the proplet the previous rule application 6.1.7 navigated to). With these values, the second proplet pattern at the rule level retrieves the second proplet shown at the Word Bank level, i.e., the motion proplet **left**. Because the rule package of Rule_2 calls Rule_2, this rule is reapplied until there are no more sequence-internal successor proplets in the Word Bank.

---

[2] The mechanism of this navigation is the same as the one illustrated in 3.4.2 for LA-think and in 3.4.3 for LA-speak.

The final proplet of a motion sequence is characterized formally by its empty next value, just as the initial proplet has an empty prev value (6.1.2, 6.1.3, 6.1.4, 6.1.7). The Word Bank proplets activated by the navigation of the LA-act1 grammar are used as blueprints for the agent's action components.

LA-act1 is simple because the concatenation between proplets is limited to the continuation attributes prev and next. Also, building the hardware for the core values *red, green, blue, strght, left,* and *right* should not be too difficult. The LA-act1 system illustrates the basic DBS constructs listed in 1.1.2.

LA-act1 resembles the propositional calculus of Symbolic Logic in that both model *coordination*.[3] They differ, however, in that propositional calculus is designed to define the truth conditions of formulas like $(p \wedge q) \vee r$, while LA-act1 is designed to model the behavior of an agent in the form of stimulus-response sequences. Accordingly, the constants and variables of standard bivalent propositional calculus have only two semantic values, namely true (T, 1) and false (F, 0), while the semantic values of LA-act1 proplets comprise an open number of recognition and action procedures, such as those defined in 6.1.5.

6.2 Guided Patterns to Expand a Fixed Behavior Repertoire

Conceptually, the interaction between a Word Bank and an LA-act grammar may be compared to the interaction between a (vinyl) record and a record player.[4] However, while the signals on a record are stored in a time-linear sequence in the record groove, the temporal dimension is recoded in a Word Bank by means of certain proplet-internal values, allowing the signals to be stored in alphabetically ordered token lines. The temporal order of a sequence is realized by the rule applications of the LA-act grammar, which drive the focus along the semantic relations between proplets, coded in the continuation attribute (here next) and the book-keeping attribute[5] (here prn).

Providing a fixed behavior agent with additional stimulus-response pairs requires translating them into sets of proplets (as in 6.1.2–6.1.4), which are stored in the agent's Word Bank. Thus, it is not an extension of the LA-act rule

---

[3] See Hausser 2003 for a reconstruction of propositional calculus in DBS, with special attention to Boolean Satisfiability (SAT).

[4] This two-level structure is one of several differences between the formalism of LA-grammar and formalisms based on possible substitution, such as Phrase Structure Grammar, including Finite State Automata. For a comparative complexity analysis of LA-grammar and PS-grammar, see FoCL'99, Chaps. 8–12. For an analysis of the relation between LA-grammar and FSAs, see CoL'89, Sect. 8.2.

[5] For a more detailed explanation of the different attribute kinds of a proplet, see NLC'06, Sect. 4.1.

system (record player) which must bear the burden of handling additional motion patterns, but an extension of the data stored in the Word Bank (records).[6]

The addition of new stimulus-response pairs may be programmed directly into the Word Bank, but it may also be "taught" to a fixed behavior agent by showing it the stimulus and guiding it through the response – like a tennis coach taking a player's arm and guiding it through a particular strike motion.[7]

As a formal example of what we call the *guided pattern* method in DBS, consider providing the fixed behavior agent described in the previous section with the new stimulus sequence **red green** and guiding it through the new motion sequence **strght, left, left, left, left,** as shown graphically below:

6.2.1 NEW PATTERN FOR A FIXED BEHAVIOR AGENT

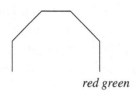

red green

This new motion pattern may be implemented by adding the following content to the agent's Word Bank:

6.2.2 CODING MOTION TRIGGERED BY red green AS SET OF PROPLETS

$$
\begin{bmatrix} \text{rec: red} \\ \text{prev:} \\ \text{next: green} \\ \text{prn: } q_1 \end{bmatrix}
\begin{bmatrix} \text{rec: green} \\ \text{prev: red} \\ \text{next: strght} \\ \text{prn: } q_2 \end{bmatrix}
\begin{bmatrix} \text{act: strght} \\ \text{prev: green} \\ \text{next: left} \\ \text{prn: } q_3 \end{bmatrix}
\begin{bmatrix} \text{act: left} \\ \text{prev: strght} \\ \text{next: left} \\ \text{prn: } q_4 \end{bmatrix}
\begin{bmatrix} \text{act: left} \\ \text{prev: left} \\ \text{next: left} \\ \text{prn: } q_5 \end{bmatrix}
\begin{bmatrix} \text{act: left} \\ \text{prev: left} \\ \text{next: left} \\ \text{prn: } q_6 \end{bmatrix}
\begin{bmatrix} \text{act: left} \\ \text{prev: left} \\ \text{next:} \\ \text{prn: } q_7 \end{bmatrix}
$$

To automatically derive the new pattern by guiding the agent through the motion shown in 6.2.1, the "record player" system of an LA-act grammar must be complemented by a co-designed "recorder" system of an LA-rec grammar (for recognition). For this the agent must be extended from being able to recognize the stimuli to also being able to recognize the elements of the guided responses.

The first step is a definition of lexical proplets, such as the following:

6.2.3 LEXICAL PROPLETS OF AN EXTENDED FIXED BEHAVIOR AGENT

$$
\begin{bmatrix} \text{rec: red} \\ \text{prev:} \\ \text{next:} \\ \text{prn:} \end{bmatrix}
\begin{bmatrix} \text{rec: green} \\ \text{prev:} \\ \text{next:} \\ \text{prn:} \end{bmatrix}
\begin{bmatrix} \text{rec: blue} \\ \text{prev:} \\ \text{next:} \\ \text{prn:} \end{bmatrix}
\begin{bmatrix} \text{act: strght} \\ \text{prev:} \\ \text{next:} \\ \text{prn:} \end{bmatrix}
\begin{bmatrix} \text{act: left} \\ \text{prev:} \\ \text{next:} \\ \text{prn:} \end{bmatrix}
\begin{bmatrix} \text{act: right} \\ \text{prev:} \\ \text{next:} \\ \text{prn:} \end{bmatrix}
$$

---

[6] An alternative model for controlling behavior is the *goal stacks* or *goal trees* in the framework of Anderson's ACT-* theory (Corbett et al. 1988).

[7] In machine learning, this would be an example of *learning from instruction.* Cf. Mitchell (1997).

These proplets are lexical because only the core attributes have values, here the attributes **rec** (for recognition) and **act** (for action) .

Once the fixed behavior agent has been extended to recognize the action steps of certain guided responses, the lexical lookup triggered by guiding the agent through the steps of 6.2.1 will result in the following sequence:

6.2.4 RECOGNITION AND LEXICAL LOOKUP OF MOTION PATTERN 6.2.1

$$
\begin{bmatrix} \text{rec: red} \\ \text{prev:} \\ \text{next:} \\ \text{prn: } q_1 \end{bmatrix}
\begin{bmatrix} \text{rec: green} \\ \text{prev:} \\ \text{next:} \\ \text{prn: } q_2 \end{bmatrix}
\begin{bmatrix} \text{act: strght} \\ \text{prev:} \\ \text{next:} \\ \text{prn: } q_3 \end{bmatrix}
\begin{bmatrix} \text{act: left} \\ \text{prev:} \\ \text{next:} \\ \text{prn: } q_4 \end{bmatrix}
\begin{bmatrix} \text{act: left} \\ \text{prev:} \\ \text{next:} \\ \text{prn: } q_5 \end{bmatrix}
\begin{bmatrix} \text{act: left} \\ \text{prev:} \\ \text{next:} \\ \text{prn: } q_6 \end{bmatrix}
\begin{bmatrix} \text{act: left} \\ \text{prev:} \\ \text{next:} \\ \text{prn: } q_7 \end{bmatrix}
$$

To complete this sequence into the proplets of 6.2.2, the core values must be cross-copied to the **prev** and **next** slots of adjacent proplets. This is done by the following LA-rec1 grammar, which uses the same variable definition as LA-act1 (6.1.5):

6.2.5 RULE SYSTEM OF THE LA-REC1 GRAMMAR

$ST_S =_{def} \{ (\, [\text{rec: } T_n] \, \{ \text{Rule\_0, Rule\_1} \}) \}$

Rule_0　　　{Rule_0, Rule_1}

$$
\begin{bmatrix} \text{rec: } T_n \\ \text{next:} \\ \text{prn: } K_i \end{bmatrix}
\begin{bmatrix} \text{rec: } T_{n+1} \\ \text{prev:} \\ \text{prn: } K_{i+1} \end{bmatrix}
\Rightarrow
\begin{bmatrix} \text{rec: } T_n \\ \text{next: } T_{n+1} \\ \text{prn: } K_i \end{bmatrix}
\begin{bmatrix} \text{rec: } T_{n+1} \\ \text{prev: } T_n \\ \text{prn: } K_{i+1} \end{bmatrix}
$$

Rule_1　　　{Rule_2}

$$
\begin{bmatrix} \text{rec: } T_n \\ \text{next:} \\ \text{prn: } K_i \end{bmatrix}
\begin{bmatrix} \text{act: M1} \\ \text{prev:} \\ \text{prn: } K_{i+1} \end{bmatrix}
\Rightarrow
\begin{bmatrix} \text{rec: } T_n \\ \text{next: M1} \\ \text{prn: } K_i \end{bmatrix}
\begin{bmatrix} \text{act: M1} \\ \text{prev: } T_n \\ \text{prn: } K_{i+1v} \end{bmatrix}
$$

Rule_2　　　{Rule_2}

$$
\begin{bmatrix} \text{act: M1} \\ \text{next:} \\ \text{prn: } K_i \end{bmatrix}
\begin{bmatrix} \text{act: M2} \\ \text{prev:} \\ \text{prn: } K_{i+1} \end{bmatrix}
\Rightarrow
\begin{bmatrix} \text{act: M1} \\ \text{next: M2} \\ \text{prn: } K_i \end{bmatrix}
\begin{bmatrix} \text{act: M2} \\ \text{prev: M1} \\ \text{prn: } K_{i+1} \end{bmatrix}
$$

$ST_F =_{def} \{ (\, [\text{next: }] \, rp_{\text{Rule\_2}}) \}$

During a derivation, the parser ensures that all instances of **K** match the same letter value.

The LA-rec1 grammar 6.2.5 is completely general. Given the variable definition 6.1.5, it handles all possible guided patterns using the stimuli **red**, **green**, and **blue** and the response steps **strght**, **left**, and **right**. If there is no restriction on the length of the different sequences, their number is infinite.[8] Furthermore,

---

[8] In a concrete agent, the number of stimulus-response pairs actually implemented will always be finite, thus requiring only a fixed amount of memory.

the LA-rec1 grammar will also handle any extensions of the variable definition (requiring concomitant extensions of (i) the agent's hardware for recognition and/or action and (ii) the memory, but not of the LA-rec1 grammar).

LA-act and LA-rec grammars fit into the component structure of diagram 4.5.3. Both receive input from the I/O component, and both are part of the rule component.

An LA-act grammar receives a stimulus as input and selectively activates proplets in the Word Bank, copies of which are passed as blueprints to the I/O component for realization.[9] An LA-rec grammar receives guided patterns from the I/O component, uses them for lexical lookup in the Word Bank, and connects the sequence of lexical proplets into stimulus-response pairs[10] – ready to be activated by the co-designed LA-act grammar.

Intuitively, the meaning of the concepts **red, green, blue, strgt, left,** and **right** equals that of the English words, used as placeholders. Procedurally, the concepts have straightforward implementations in terms of artificial vision (measuring electromagnetic frequencies) and locomotion. The intuitive and the procedural characterization of concepts are equally necessary to ensure that the procedures are attached to the correct placeholders used as handles.[11]

The agent-internal representation of the external world by means of these concepts is extremely sparse – in concord with the tenets of subsumption architecture in robotics (Brooks 1991). For example, a fixed behavior robot trying to execute the sequence 6.2.2 in rough terrain could not possibly model this terrain, given the limits of its recognition. Instead the realization of the motion steps is left to a loosely coupled, massively parallel, analog walking machine as described by Tilden and Hasslacher (1994). This machine is *subsymbolic* because it uses "digital pulse trains ... for motor drive and control" (op. cit.).

For higher-level reasoning, such subsymbolic procedures must be related to symbolic ones. Thereby, nouvelle AI proceeds from the subsymbolic to the symbolic, while DBS proceeds from the symbolic to the subsymbolic. For example, the fixed behavior agent outlined above is symbolic, but assumes a subsymbolic, procedural realization of the core values in terms of the agent's elementary recognition and action procedures.

For learning, however, the relevant transition is not from the symbolic to the subsymbolic or vice versa, but from fixed behavior to adaptive behavior. In contradistinction to expanding the repertoire of a fixed behavior agent by means of guided patterns (provided by a scientist), adaptation and learning must be autonomous, driven by automatic appraisal and schema derivation.

---

[9] This is analogous to 4.6.2.

[10] This is analogous to 4.6.1.

6.3 Transition from Fixed to Adaptive Behavior

Evolving a software system of fixed behavior into one of adaptive behavior requires a number of rather obvious extensions:

6.3.1 EXTENSIONS REQUIRED BY AN ADAPTIVE BEHAVIOR AGENT

1. **Writable memory**
   In order to record individual recognition action episodes, the agent's non-writable memory must be complemented with a writable memory.

2. **Decoupling of recognition and action**
   The agent must be capable of recognition without having to perform the associated fixed behavior action (*recognition per se*), just as there must be action triggered by reasoning rather than by a fixed behavior stimulus.

3. **Unknowns**
   The agent must be able to recognize and store unknowns consisting of previously unencountered constellations of available recognition elements.[12]

4. **Appraisal**
   In order to learn from past experiences, the agent must be able to evaluate the implication of recognitions and the outcome of actions.[13]

5. **Automatic schema derivation**
   In order to generalize over similar constellations, the agent must be capable of automatic schema derivation (Sect. 6.4).

It follows from (1) that fixed behavior and adaptive behavior may be distinguished in terms of a non-writable vs. a writable memory. For fixed behavior, a non-writable memory is sufficient: once a stimulus-response pattern has been coded, no additional memory is needed for repeating it. An adaptive behavior agent, however, requires an additional writable memory for recording individual episodes of recognition and action in the order of their arrival at the *now front*.[14]

---

[11] In this particular respect, the difference between natural languages boils down to the use of different placeholders (handles) in the form of language-dependent surfaces.

[12] An example from vision is *recognition by components* (RBC) based on geons, proposed by Biederman (1987). Cf. L&I'05: "Memory-Based Pattern Completion in Database Semantics."

[13] Ekman 2003, p. 31, speaks of the automatic appraisal mechanism in the "emotion database" of a cognitive agent. In DBS, appraisal (or evaluation) is integrated into the proplets in the form of values for an additional appraisal attribute. Thus, the only database needed for appraisal is the Word Bank.

[14] In nature, practically all living beings, starting with the protozoa, are capable of some form of associative learning (classical conditioning), and must therefore have some writable memory. The guided pattern method for extending a fixed behavior repertoire (Sect. 6.2) also requires a certain amount of

Consider, for example, the fixed behavior agent described in Sect. 6.1. Its Word Bank has token lines with the owner proplets *blue, green, left, red, right,* and *strght* in alphabetical order. By storing fixed behavior proplets to the right of the owner proplets and adding "episodic" proplets to the left, the writable and the non-writable memory are combined using the same owner proplets:

6.3.2 ARRANGEMENT OF WRITABLE AND NON-WRITABLE MEMORY

writable memory			*now front*	*owner proplets*	*non-writable memory*

$$\begin{bmatrix} \text{rec: blue} \\ \text{prev:} \\ \text{next: strght} \\ \text{prn: 21} \end{bmatrix} \begin{bmatrix} \text{rec: blue} \\ \text{prev:} \\ \text{next: strght} \\ \text{prn: 37} \end{bmatrix} \begin{bmatrix} \text{rec: blue} \\ \text{prev:} \\ \text{next: strght} \\ \text{prn: 55} \end{bmatrix} \cdots \quad [\text{blue}] \qquad \begin{bmatrix} \text{rec: blue} \\ \text{prev:} \\ \text{next: strght} \\ \text{prn: } z_1 \end{bmatrix}$$

$$\begin{bmatrix} \text{rec: green} \\ \text{prev:} \\ \text{next: strght} \\ \text{prn: 14} \end{bmatrix} \begin{bmatrix} \text{rec: green} \\ \text{prev:} \\ \text{next: strght} \\ \text{prn: 38} \end{bmatrix} \begin{bmatrix} \text{rec: green} \\ \text{prev:} \\ \text{next: strght} \\ \text{prn: 42} \end{bmatrix} \cdots \quad [\text{green}] \qquad \begin{bmatrix} \text{rec: green} \\ \text{prev:} \\ \text{next: strght} \\ \text{prn: } y_1 \end{bmatrix}$$

To be concise, only the first two token lines of the Word Bank in question are shown. A complete Word Bank would all 21 proplets of 6.1.2, 6.1.3, and 6.1.4, sorted into the 6 token lines of the non-writable memory to the right of the owner proplets.

The writable memory is filled by copying[15] each proplet of a sequence performance to the left of the corresponding owner proplet (i.e., to the *now front*). In such an episodic proplet, the alphabetic prn values, e.g., z_1, are replaced by incremented numerical values, e.g., 55. The episodic proplets in each token line are in the temporal order of their arrival, as reflected by the increasing prn values, e.g., 21, 37, 55, ...

The extension illustrated in 6.3.2 keeps track of how often a fixed behavior pattern was activated in the past and in what order the activations occurred, but there are no recognitions per se. In order to allow the agent to observe and remember without having to execute an associated fixed behavior, there must also be the possibility of decoupling recognition and action (cf. 2 in 6.3.1).[16]

Decoupled elementary items may be derived automatically from any given fixed behavior. Consider, for example, an artificial cognitive agent which tries to hide when it recognizes a green light, but continues feeding when it sees a red light. This fixed behavior may be formulated as the following two R/E one-step inference chains (5.2.3):

---

writable memory. In other words, the strictly non-writable memory of our fixed behavior agent is a convenient theoretical construct, applied in the practical software construction of artificial agents.

[15] This description is conceptual and does not prescribe technical details of the actual implementation.

[16] Instances of non-decoupled behavior are the reflexes.

6.3.3 TWO SIMPLE R/E ONE-STEP INFERENCE CHAINS

R/E [rec: red square]   K   cm/exec [act: feed]   K+1
R/E [rec: green circle]   K   cm/exec [act: hide]   K+1

Automatic decoupling consists of disassembling the stimulus-response pairs of the agent's fixed behavior into their elementary parts. For example, the disassembling of 6.3.3 results in the following elements, each a separate and manageable task of software and hardware engineering:

6.3.4 DECOUPLED RECOGNITIONS AND ACTIONS

rec: red   act: hide
rec: green   act: feed
rec: square
rec: circle

Any execution of the fixed behavior inferences 6.3.3 is copied into the agent's writable memory in the form of decoupled recognitions and actions as defined in 6.3.4. Thereby, they receive incremented numerical **prn** values, resulting in their reconnection in terms of their temporal sequence (coordination).

Let us assume that the combinations **red square** and **green circle** are knowns, while the alternative combinations **green square** and **red circle** are unknowns (cf. 3 in 6.3.1). When faced with such unknowns, two kinds of actions are available to the agent, namely **act: feed** and **act: hide**. Without additional assumptions, the choice between these options is random.

For appraisal (cf. 4 in 6.3.1), let us add two more decoupled internal recognitions to 6.3.4, namely [rec: good] and [rec: bad]. When the agent is faced with an unknown, e.g., **red circle**, there are the following possibilities:

6.3.5 POSSIBLE CONSTELLATIONS WHEN FACED WITH AN UNKNOWN

1 rec: red circle   act: hide   rec: good
2 rec: red circle   act: hide   rec: bad
3 rec: red circle   act: feed   rec: good
4 rec: red circle   act: feed   rec: bad

For example, if the agent recognizes a red circle, chooses to hide, and evaluates the decision as bad (because it interrupts feeding unnecessarily), as in 2 of 6.3.5, the next[17] encounter of a red circle will not be at random, based on the following general inference rule:

---

[17] In real life, agents usually do not change their behavior after the first negative experience. This makes functional sense insofar as instances of the same behavior pattern may call forth different consequences from the external environment. However, the worse the consequence of an initial random behavior, the more likely the avoidance of that behavior in a second encounter.

6.3.6 Consequence inference for negative experience (CIN)

rec: α act: β rec: bad   **csq**   rec: α act: no β

Like all DBS inference rules, CIN derives new content (consequent) from
given content (antecedent), here using the connective **csq** (for consequence).
Unlike the inference rules 5.2.3, 5.2.5, 5.3.3, and 5.3.5, the consequent of 6.3.5
is not connected to the antecedent by one or more addresses. This is because
the consequent may apply to an instantiation of α different from that in the
antecedent – though identity at the content level is not precluded.

In a second encounter with a red circle, the antecedent of inference 6.3.6
matches the negative experience 2 in 6.3.5 with a previous unknown. Based
on the alternative blueprint for action provided by the consequent of the CIN
inference, this second encounter will not result in hiding, thus preventing an
unnecessary interruption of the agent's feeding activity. Applying CIN to the
negative experience 4 of 6.3.5 has the opposite effect, causing the agent to hide
on next encountering a red circle.

The positive experiences 1 and 3 of 6.3.5, in contrast, reenforce the action
initially chosen at random, based on the following inference:

6.3.7 Consequence inference for positive experience (CIP)

rec: α act: β rec: good   **csq**   rec: α act: β

In this way, adaptive behavior derives new action patterns in the agent's
writable memory which serve to maintain the agent's balance.[18]

6.4 Upscaling from Coordination to Functor-Argument

The combination of old elementary "knowns" into a new complex "unknown"
(cf. 3 in 6.3.1), such as red circle in 6.3.5, goes beyond the method of using
rec and act as the only core attributes of proplets, practiced for simplicity
in the definitions of LA-act1 (6.1.6) and LA-rec1 (6.2.5). For example, given
that red circle is a functor-argument, with the adjective red modifying the
noun circle, we need to refine the LA-act1 and LA-rec1 grammars to handle
functor-argument in addition to coordination.

This refinement resembles the transition from propositional calculus to first-
order predicate calculus in Symbolic Logic. Symbolic Logic constructs the
transition by building formulas of predicate calculus from formulas of propositional calculus, using the additional constructs of functors, arguments, variables, and quantifiers. Consider the following example:

6.4.1 USE OF PROPOSITIONAL CALCULUS IN PREDICATE CALCULUS

$$[p \wedge q] \implies \exists x[red(x) \wedge circle(x)]$$

The unanalyzed propositions **p** and **q** preceding the arrow are replaced by **red(x)** and **circle(x)** in the formula of predicate calculus following the arrow.[19] The quantifier $\exists x$ horizontally binds the two instances of the variable x. The formula of predicate calculus inherits from the formula of propositional calculus the connective $\wedge$ (including its truth table) and the bracketing structure surrounding it.

DBS, in contrast, accommodates the extension to functor-argument within the structural means of the proplet format, as shown below:

6.4.2 INTEGRATING FUNCTOR-ARGUMENT IN DBS

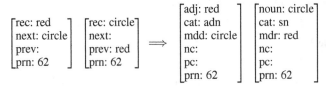

The two proplets preceding the arrow treat **red** and **circle** as elementary recognitions,[20] indicated by the core attribute **rec**. They are connected by the values of the continuation attributes **next** and **prev**.

In the two proplets following the arrow, the core attributes **rec** are revised into the more differentiated attributes **adj** and **noun**, characterizing the parts of speech. The continuation attributes **prev** and **next** reappear as **nc** (for next conjunct) and **pc** (for previous conjunct). They are used for the intrapropositional coordination of adjectives, as in **beautiful, young, intelligent** (9.6.6, 9.6.7), and nouns, as in **the man, the woman, and the child** (cf. 9.2.5; NLC'06, Sect. 8.2).

The common **prn** value of the *red* and *circle* proplets, here **62**, indicates their belonging to the same proposition, and assumes the binding function of the quantifier in 6.4.1. The kind of semantic relation between the two proplets is established by the continuation attributes, here **mdd** and **mdr**, and specified further by their values, **circle** and **red**, respectively.

---

[18] In machine learning, CIN and CIP would be examples of *learning from experience.*

[19] Even though **red circle** is linguistically a phrasal noun and not a sentence, it is treated in 6.4.1 as a proposition insofar as **red(x)** and **circle(x)** each denote a truth value.

[20] We refrain from calling an elementary recognition or action a "proposition." A set of proplets with a common **prn** value is called a proposition only if it contains a verb – which presupposes functor-argument.

In the proplets preceding the arrow, the semantic relation is coordination, expressed by the attributes **prev** and **next**. In the proplets following the arrow, the semantic relation is functor-argument in the form of modifier|modified, expressed by the attributes **mdr** and **mdd**. The content *red circle* is characterized as a phrasal noun, and may be turned into a proposition by adding a suitable verb.

In addition to the upscaling from coordination only to coordination plus functor-argument there is the upscaling from nonlanguage content to language content. The latter is treated as another refinement of the basic proplet structure. This is shown by the following Word Bank example containing the content *red circles* as coreferent language proplets and nonlanguage proplets:

6.4.3 CODING NONLANGUAGE AND LANGUAGE CONTENT ALIKE

member proplets	now front	owner proplets
...		...

$$
\begin{bmatrix} \text{sur:} \\ \text{noun: circle} \\ \text{cat: pnp} \\ \text{sem: pl sel} \\ \text{fnc:} \\ \text{mdr: red} \\ \text{nc:} \\ \text{pc:} \\ \text{prn: 37} \end{bmatrix}
\quad \dots \quad
\begin{bmatrix} \text{sur: circles} \\ \text{noun: (circle 37)} \\ \text{cat: pnp} \\ \text{sem: pl sel} \\ \text{fnc:} \\ \text{mdr: (red 37)} \\ \text{nc:} \\ \text{pc:} \\ \text{prn: 62} \end{bmatrix}
\qquad
[\text{core: circle}]
$$

$$
\begin{bmatrix} \text{sur:} \\ \text{adj: red} \\ \text{cat: adn} \\ \text{sem: psv} \\ \text{mdd: circle} \\ \text{nc:} \\ \text{pc:} \\ \text{prn: 37} \end{bmatrix}
\quad \dots \quad
\begin{bmatrix} \text{sur: red} \\ \text{adj: (red 37)} \\ \text{cat: adn} \\ \text{sem: psv} \\ \text{mdd: (circle 37)} \\ \text{nc:} \\ \text{pc:} \\ \text{prn: 62} \end{bmatrix}
\qquad
[\text{core: red}]
$$

In contradistinction to the "vertical" matching between the language and the context component shown in 4.3.2, the above example treats reference as a "horizontal" matching between language and context proplets within token lines, in accordance with the refined component structure of a cognitive agent presented in 4.5.3 and 4.5.4.

The difference between the language and nonlanguage proplets concerns their respective **sur** attributes. The language proplets with the **prn** value **62** have the nonempty **sur** values **red** and **circles**, whereas the corresponding **sur** slots in the context proplets with the **prn** value **37** have the value NIL, represented by empty space.

The order of content and language proplets in a token line is determined solely by the order of their arrival. In other words, they may be freely mixed.

Which language proplet refers to which content proplet is coded by addresses defined as core values. An original may have many coreferent addresses. It holds that an address always follows the original in a token line.

For example, the coreference between the language and the context proplets is coded by the address values (red 37) and (circle 37) in the language proplets. The core value of the language proplets is a concept type corresponding to the owner proplet, while the core value of the context proplets is usually a concept token (4.3.3).

The determiner properties of the quantifiers $\exists x$ (for some) and $\forall x$ (for all) of predicate calculus are recoded in DBS as values of the sem attribute, i.e., [sem: pl sel] for some and [sem: pl exh] for all. The values pl, sel, and exh stand for plural, selective, and exhaustive, respectively. For a set-theoretical characterization of these values as well as the definition of other determiners like a(n), every, and the, see NLC'06, 6.2.9.

In the hear mode, the constellation of coreferential nonlanguage and language proplets shown in 6.4.3 comes about by (i) parsing the surface some red circles into connected *red* and *circle* proplets, (ii) turning these proplets temporarily into patterns by using a variable as their prn value, (iii) using the patterns to search their token lines for matching referents, (iv) replacing their core values by suitable addresses, and (v) storing them at the *now front* of the Word Bank. In the speak mode, the context proplets are (i) copied to the *now front*, (ii) provided with a new (current) prn value and addresses for the core values, and (iii) are passed to LA-speak for language production.

The uniform coding of language and nonlanguage proplets is in concord with the Humboldt-Sapir-Whorf hypothesis (linguistic relativism, cf. Sect. 3.6), according to which a natural language influences the thought of its speaker-hearers. DBS adopts also the reverse direction by treating natural language expressions as a direct reflection of thought. This has the following impact on the computational treatment of reference in DBS:

6.4.4 THE FIFTH MECHANISM OF COMMUNICATION (MOC-5)

Exporting coordination and functor-argument from language content to nonlanguage (context) content, coded uniformly as sets of proplets, allows us to model *reference* as a pattern matching between language content and context content in the hear mode, and between context content and language content in the speak mode.

In DBS, linguistic relativism is counterbalanced by the universals of natural language listed in 1.1.1–1.1.4.

6.5 Schema Derivation and Hierarchy Inferencing

By using (i) a common **prn** value in lieu of the variable-binding function of the logical quantifiers and (ii) **sem** values in lieu of the determiner function of $\exists x$ and $\forall x$, the use of variables in DBS may be limited to *pattern proplets*, in contradistinction to *content proplets*, which by definition may not contain any variables. Pattern proplets are combined into schemata for the purpose of matching corresponding contents.

A schema may be derived from a content by replacing constants with variables (simultaneous substitution). The set of content proplets matched by a schema is called its *yield*. The yield of a schema relative to a given Word Bank may be controlled precisely by two complementary methods. One is by the choice and number of constants in a content which are replaced by variables. The other is by variable restrictions.

The use of restricted variables allows us to convert any content into strictly equivalent schemata and any schema into strictly equivalent contents. As an example of the content-schema conversion consider the content corresponding to **Every child slept. Fido snored.**, i.e., the coordination of two propositions, each with a subject-verb construction, one with a phrasal noun, the other with a proper name as subject.[21]

6.5.1 CONVERTING A CONTENT INTO AN EQUIVALENT SCHEMA

content

noun: child		verb: sleep		noun: Fido		verb: snore	
cat: snp		cat: decl		cat: nm		cat: decl	
sem: pl exh		sem: past		sem: animal		sem: past	
fnc: sleep		arg: child		fnc: snore		arg: Fido	
mdr:		mdr:		mdr:		mdr:	
nc:		nc: (snore 27)		nc:		nc:	
pc:		pc:		pc:		pc: (sleep 26)	
prn: 26		prn: 26		prn: 27		prn: 27	

$\Longleftrightarrow$

schema

noun: α	verb: β	noun: γ	verb: δ
cat: snp	cat: decl	cat: nm	cat: decl
sem: pl exh	sem: past	sem: animal	sem: past
fnc: β	arg: α	fnc: δ	arg: γ
mdr:	mdr:	mdr:	mdr:
nc:	nc: (δ K+1)	nc:	nc:
pc:	pc:	pc:	pc: (β K)
prn: K	prn: K	prn: K+1	prn: K+1

where $\alpha \in \{child\}$, $\beta \in \{sleep\}$, $\gamma \in \{Fido\}$, $\delta \in \{snore\}$, and $K \in \{26\}$

---

[21] The fact that **every** takes a singular noun but refers to a plural set is expressed by the **cat** value snp, for singular noun phrase, and the **sem** values pl exh. The phrasal noun and the proper name proplet have the same attribute structure, but are distinguished by means of different **cat** and **sem** values. This way of treating phrasal nouns and proper names alike is much simpler than type raising in lambda calculus (Montague 1974, PTQ).

In this example, all core and **prn** values of the content are simultaneously substituted with variables in the schema (method one) and all variables are restricted to the value they replace (method two). In this way, strict equivalence between the content and the schema representation in 6.5.1 is obtained.

The yield of a schema may be enlarged by adding values to the restriction set of variables. Consider the following example of a schema matching the content corresponding to **Every child slept**, but with extended variable restrictions.

6.5.2 CONVERTING A SCHEMA INTO EQUIVALENT CONTENTS

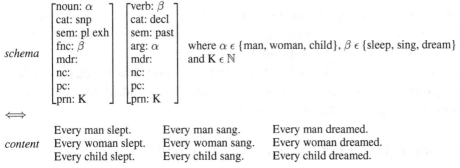

$$schema \quad \begin{bmatrix} \text{noun: } \alpha \\ \text{cat: snp} \\ \text{sem: pl exh} \\ \text{fnc: } \beta \\ \text{mdr:} \\ \text{nc:} \\ \text{pc:} \\ \text{prn: K} \end{bmatrix} \begin{bmatrix} \text{verb: } \beta \\ \text{cat: decl} \\ \text{sem: past} \\ \text{arg: } \alpha \\ \text{mdr:} \\ \text{nc:} \\ \text{pc:} \\ \text{prn: K} \end{bmatrix} \quad \text{where } \alpha \in \{\text{man, woman, child}\}, \beta \in \{\text{sleep, sing, dream}\} \\ \text{and } K \in \mathbb{N}$$

$\Longleftrightarrow$

content	Every man slept.	Every man sang.	Every man dreamed.
	Every woman slept.	Every woman sang.	Every woman dreamed.
	Every child slept.	Every child sang.	Every child dreamed.

The contents are generated from the schema by systematically replacing the variables α and β with elements of their restriction sets. Even if no restriction set is specified, the yield of a schema relative to an actual Word Bank will be finite.

A schema may be derived automatically from any set of partially overlapping contents. Consider the following example:

6.5.3 SET OF CONTENTS WITH PARTIAL OVERLAP

Julia eats an apple	John eats an apple	Suzy eats an apple	Bill eats an apple
Julia eats a pear	John eats a pear	Suzy eats a pear	Bill eats a pear
Julia eats a salad	John eats a salad	Suzy eats a salad	Bill eats a salad
Julia eats a steak	John eats a steak	Suzy eats a steak	Bill eats a steak

Of these 16 propositions, each contains the proplet *eat*, while the proplets *Julia, John, Suzy*, and *Bill* occur four times as subject and the proplets *apple, pear, salad*, and *steak* occur four times as object. Based on these repetitions, the propositions may be summarized as the following schema:

6.5.4 SUMMARIZING THE SET 6.5.3 AS A SCHEMA

$$\begin{bmatrix} \text{noun: } \alpha \\ \text{fnc: eat} \\ \text{prn: K} \end{bmatrix} \begin{bmatrix} \text{verb: eat} \\ \text{arg: } \alpha \, \beta \\ \text{prn: K} \end{bmatrix} \begin{bmatrix} \text{noun: } \beta \\ \text{fnc: eat} \\ \text{prn: K} \end{bmatrix} .$$

where $\alpha \in \{\text{Julia, John, Suzy, Bill}\}$ and $\beta \in \{\text{apple, pear, salad, steak}\}$

Due to the restriction on the variables α and β, 6.5.4 is strictly equivalent to 6.5.3. From a linguistic point of view, 6.5.4 may be seen as the valency schema (Herbst et al. 2004) or lexical frame of the transitive verb *eat*. The restriction sets of the variables α and β may be established automatically by parsing a corpus: all subjects of *eat* actually occurring in the corpus are written into the restriction set of α and all objects are written into the restriction set of β (Sect. 8.5).

One of the many results of automatic schema derivation is the *is-a* hierarchy (subsumptive containment hierarchy) familiar from knowledge representation and object-oriented programming. As a subclass relation, the *is-a* hierarchy is a prime example of the substitution approach because it is motivated more directly and more obviously than the constituent structure trees of Phrase Structure Grammar and Categorial Grammar.

The subclass relation between **food** as the hypernym and the set **apple, pear, salad, steak** as the instantiation may be coded by extending the restriction technique from variables to certain constants:

6.5.5 CODING THE SUBCLASS RELATION FOR **food**

$$\begin{bmatrix} \text{noun: food} \\ \text{fnc: } \beta \\ \text{prn: K} \end{bmatrix}$$

where *food* ϵ {apple, pear, salad, steak}

The hypernym concept *food* may serve as the literal meaning of the word **food** in English, **aliment** in French, **Nahrung** in German, and so on.

The derivation of a semantic hierarchy, as illustrated in 6.5.3–6.5.5, is empirically adequate if the resulting class containing the instantiations corresponds to that of the surrounding humans. For example, if the artificial agent observes humans to habitually (frequency) eat müsli, the restriction list of *food* must be adjusted correspondingly.[22] Furthermore, the language surface, e.g., **aliment**, chosen by the artificial agent for the hypernym concept, e.g., *food*, must correspond to that of the natural language in use.

Implicit in schema 6.5.5 is the following tree structure:

6.5.6 REPRESENTING THE SEMANTIC HIERARCHY 6.5.5 AS A TREE

---

[22] This resembles the establishment of inductive inferences in logic, though based on individual agents.

Just as such a tree requires some tree-walking algorithm to get from a higher node to a lower node or from a lower node to a higher node, a schema like 6.5.5 requires DBS inferences to utilize the content of the hierarchy for reasoning. For example, an inference for downward traversal should allow the agent to infer that *food* may be instantiated by apples, pairs, salad, or steak, while an inference for upward traversal should infer that an apple, for example, instantiates *food*.[23]

Given that semantic hierarchy relations abound in the lexicon, we derive the inferences for their upward and downward traversal automatically by means of a meta-inference which takes a schema like 6.5.5 as input and derives the associated inferences for upward and downward traversal:

6.5.7 META-INFERENCE DERIVING **down** AND **up** INFERENCES

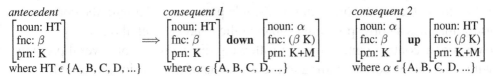

The variable HT stands for a higher-term constant, e.g., *food*, while the variables A, B, C, D, …, stand for the instantiation constants, e.g., apple, pear, salad, steak. Other examples of higher term constants are *mammal*, which is instantiated by dogs, cats, and mice; *vehicle*, which is instantiated by cars, trucks, and buses; *water*, which is instantiated by ocean, lake, and river; and so on. They may all be derived automatically in the same way as is the higher term *food*, as shown in 6.5.3 – 6.5.5.

Application of the meta-inference 6.5.7 to the *food* proplet 6.5.5 results in the following inference for downward traversal:

6.5.8 APPLYING META-INFERENCE 6.5.7 TO DERIVE **down** INFERENCE

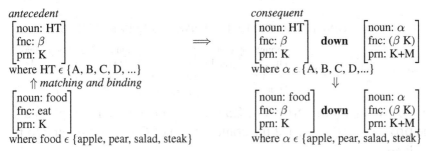

While meta-inferences are applied to pattern proplets with restricted higher

---

[23] Hypernyms in different agents may vary, depending on personal preferences.

terms as core values, the resulting inferences are applied to content proplets, i.e., proplets without any variables.

For example, if the content **Julia is looking for food** is activated, the inference for downward traversal derived in 6.5.8 may be applied to it as follows:

6.5.9 APPLYING INFERENCE FOR DOWNWARD TRAVERSAL

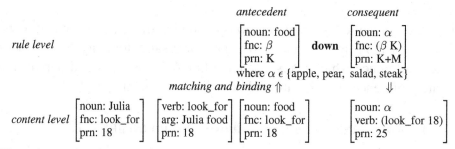

When the antecedent, consisting of a single pattern proplet with the core value **food**, matches a corresponding proplet at the content level, the consequent derives a new content containing the following disjunction[24] of several proplets with core values corresponding to the elements of the restriction set of α:

6.5.10 OUTPUT DISJUNCTION OF THE DOWNWARD INFERENCE 6.5.8

$$
\begin{bmatrix}
\text{noun: apple } or \\
\text{fnc: (look\_for 18)} \\
\text{nc: pear} \\
\text{prn: 25}
\end{bmatrix}
\begin{bmatrix}
\text{noun: pear} \\
\text{pc: apple} \\
\text{nc: salad} \\
\text{prn: 25}
\end{bmatrix}
\begin{bmatrix}
\text{noun: salad} \\
\text{pc: pear} \\
\text{nc: steak} \\
\text{prn: 25}
\end{bmatrix}
\begin{bmatrix}
\text{noun: steak} \\
\text{pc: salad} \\
\text{nc:} \\
\text{prn: 25}
\end{bmatrix}
$$

The proplets of the output disjunction are concatenated by the **pc** (for previous conjunct) and **nc** (for next conjunct) features, and have the new **prn** value **25**. They are related to the original proposition by the pointer address (**look_for 18**) serving as the **fnc** value of the first disjunct. The output disjunction may be completed automatically into the new proposition **Julia looks_for apple or pear or salad or steak**, represented as follows:

6.5.11 PROPOSITION RESULTING FROM DOWNWARD INFERENCE 6.5.9

$$
\begin{bmatrix}
\text{noun: (Julia 18)} \\
\text{fnc: (look\_for 18)} \\
\text{prn: 25}
\end{bmatrix}
\begin{bmatrix}
\text{verb: (look\_for 18)} \\
\text{arg: (Julia 18) apple } or \\
\text{prn: 25}
\end{bmatrix}
\begin{bmatrix}
\text{noun: apple } or \\
\text{fnc: (look\_for 18)} \\
\text{nc: pear} \\
\text{prn: 25}
\end{bmatrix}
\begin{bmatrix}
\text{noun: pear} \\
\text{pc: apple} \\
\text{nc: salad} \\
\text{prn: 25}
\end{bmatrix}
\begin{bmatrix}
\text{noun: salad} \\
\text{pc: pear} \\
\text{nc: steak} \\
\text{prn: 25}
\end{bmatrix}
\begin{bmatrix}
\text{noun: steak} \\
\text{pc: salad} \\
\text{nc:} \\
\text{prn: 25}
\end{bmatrix}
$$

This new proposition with the **prn** value **25** is derived from the input proposition with the **prn** value **18** shown at the content level of 6.5.9, and related to it by pointer values.

---

[24] See NLC'06, Chap. 8, for a detailed discussion of intrapropositional coordination such as conjunction and disjunction.

The inverse of downward traversal is the upward traversal of a semantic hierarchy. An upward inference assigns a hypernym like *food* to concepts like *salad* or *steak*. Consider the following application of the inference for upward traversal derived in 6.5.7 (*consequent 2*).

6.5.12 HIERARCHY-INFERENCE FOR UPWARD TRAVERSAL

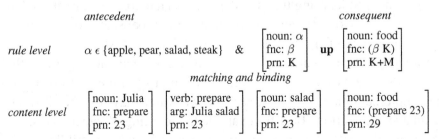

As in the downward inference 6.5.9, the antecedent of the upward inference consists of a single pattern proplet, now with the restricted variable α as the core value. Due to the use of a pointer address as the fnc value of the output there is sufficient information to complete the output proplet into the new proposition **Julia prepares food** (not shown), with the prn value **29** and the pointer proplets *(Julia 23)* and *(prepare 23)*.

The automatic derivation and restriction of schemata like 6.5.4 and 6.5.5 directly controls the automatic adaptation of the hierarchy inferences by adjusting their restriction sets. In this way, DBS fulfills the three functions which define an autonomic system: "automatically configure itself in an environment, optimize its performance using the environment and mechanisms for performance, and continually adapt to improve performance and heal itself in a changing environment" (Naphade and Smith 2009).

6.6 Natural vs. Artificial Language Learning

The mechanism of automatically deriving and adjusting DBS schemata holds at a level of abstraction which applies to natural and artificial agents alike. The simplicity of this mechanism allows us to design artificial agents as natural agents in that they adjust over time. Thereby, the following differences between natural and artificial agents do not stand in the way:

In natural agents, adjusting to a changing environment and optimizing come in two varieties, (i) the biological *adaptation* of a species in which physical abilities (hardware) and cognition (software) are co-evolved, and (ii) the *learning* of individuals, which is mostly limited to cognition (software) alone.

Adaptation and learning differ also in that they apply to different ranges of time and different media of storage (gene memory vs. brain memory).

In artificial agents, in contrast, improvement of the hardware is the work of engineers, while development of an automatically adjusting cognition is the work of software designers. Because of this division of labor between hardware and software development, the automatic adjustment of artificial agents corresponds more to learning than to adaptation. That artificial agents lack the natural inheritance from parent to child may be compensated for by copying the cognition software from one hardware model to the next.

The similarity between natural and artificial cognitive agents regarding their theoretical (declarative) structure and the dissimilarity in their practical (procedural) implementation, i.e., hardware vs. wetware, may also be illustrated with language acquisition. Take for example the learning of word forms.

For humans, this is a slow process, taking several years in childhood (first language acquisition) and in adulthood (second language acquisition). An artificial agent, in contrast, may simply be uploaded with an online dictionary and associated software for automatic word form recognition – not just for one language, but for as many languages as available or desired.

The word form analyses provided to the artificial agent in this way specify (i) the morphosyntactic properties (e.g., part of speech) formalized as *proplet shells* (cf. NLC'06, Sect. 4.1), and (ii) the core values, represented by placeholders. Thereby, the proplet shells and the core values are orthogonal to each other in the sense that (i) a given proplet shell may take different core values and (ii) a given core value may be embedded into different proplet shells.

The following example shows a proplet shell taking different core values:

6.6.1 ONE PROPLET SHELL TAKING DIFFERENT CORE VALUES

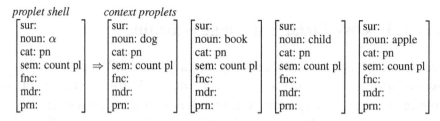

The context proplets derived from the proplet shell differ in only one value, namely that of the core attribute **noun**, which should facilitate learning.

Context proplets may be turned into language proplets by inserting the appropriate **sur** values, as in the following example for English:

6.6.2 TURNING CONTEXT PROPLETS INTO LANGUAGE PROPLETS

proplet shell   language proplets

$$
\begin{bmatrix} \text{sur: } \alpha' + x \\ \text{noun: } \alpha \\ \text{cat: pn} \\ \text{sem: count pl} \\ \text{fnc:} \\ \text{mdr:} \\ \text{prn:} \end{bmatrix} \Rightarrow
\begin{bmatrix} \text{sur: dog+s} \\ \text{noun: dog} \\ \text{cat: pn} \\ \text{sem: count pl} \\ \text{fnc:} \\ \text{mdr:} \\ \text{prn:} \end{bmatrix}
\begin{bmatrix} \text{sur: book+s} \\ \text{noun: book} \\ \text{cat: pn} \\ \text{sem: count pl} \\ \text{fnc:} \\ \text{mdr:} \\ \text{prn:} \end{bmatrix}
\begin{bmatrix} \text{sur: child+ren} \\ \text{noun: child} \\ \text{cat: pn} \\ \text{sem: count pl} \\ \text{fnc:} \\ \text{mdr:} \\ \text{prn:} \end{bmatrix}
\begin{bmatrix} \text{sur: apple+s} \\ \text{noun: apple} \\ \text{cat: pn} \\ \text{sem: count pl} \\ \text{fnc:} \\ \text{mdr:} \\ \text{prn:} \end{bmatrix}
$$

Assuming that the context proplets in 6.6.1 have been acquired already, learning the associated language proplets involves only a single value, namely that of the **sur** attribute, again facilitating learning.

Once the proplets have been acquired for one language, they may be reused for other languages, provided the lexicalization is similar.[25] The following example shows the proplets for the concept *dog* with English, French, German, and Italian surfaces:

6.6.3 TAKING SUR VALUES FROM DIFFERENT LANGUAGES

proplet shell   language proplets

$$
\begin{bmatrix} \text{sur: } \alpha' \\ \text{noun: } \alpha \\ \text{cat: sn} \\ \text{sem: count sg} \\ \text{fnc:} \\ \text{mdr:} \\ \text{prn:} \end{bmatrix} \Rightarrow
\begin{bmatrix} \text{sur: dog} \\ \text{noun: dog} \\ \text{cat: sn} \\ \text{sem: count sg} \\ \text{fnc:} \\ \text{mdr:} \\ \text{prn:} \end{bmatrix}
\begin{bmatrix} \text{sur: chien} \\ \text{noun: dog} \\ \text{cat: sn} \\ \text{sem: count sg} \\ \text{fnc:} \\ \text{mdr:} \\ \text{prn:} \end{bmatrix}
\begin{bmatrix} \text{sur: Hund} \\ \text{noun: dog} \\ \text{cat: sn} \\ \text{sem: count sg} \\ \text{fnc:} \\ \text{mdr:} \\ \text{prn:} \end{bmatrix}
\begin{bmatrix} \text{sur: cane} \\ \text{noun: dog} \\ \text{cat: sn} \\ \text{sem: count sg} \\ \text{fnc:} \\ \text{mdr:} \\ \text{prn:} \end{bmatrix}
$$

For syntactic-semantic parsing, the French, German, and Italian proplet versions will have to be complemented with the additional **cat** value **m** (for the grammatical gender masculine). This language-dependent information may be obtained from the traditional dictionaries for these languages. In addition, corpus-based information, such as domain-dependent frequency, LA-hear predecessors and successors ordered according to frequency (n-grams), semantic relations, etc., may be added to the owner proplets (Sect. 8.5).[26]

---

[25] Cf. 3.6.1; other examples of different lexicalizations are (i) German Traumreise (literally *dream journey*), which has been translated into American English as dream vacation and into French as voyage des rêves, (ii) English horseshoe, which translates into German as Hufeisen (literally *hoof iron*) and into French as fer à cheval (literally *iron for horse*), and (iii) French ralenti, which translates into English as slow motion and into German as Zeitlupe.

[26] Automatic word form recognition (based on a lexicon and rules) provides a more accurate frequency analysis of a corpus, for example, than part-of-speech tagging (based on statistical transition likelihoods from one word form to the next in a corpus). Unlike automatic word form recognition, part-of-speech tagging does not relate surface values such as learn, learns, learned, learning, and swim, swims, swam, swum, swimming, to their base forms (core values), i.e., *learn* and *swim*, respectively. Therefore, the rule-based and the statistical approach lead to substantially different frequency distribution results. For an evaluation of the CLAWS4 tagging analysis of the Britisch National Corpus (BNC), see FoCL'99, Sect. 5.5.

The second orthogonal relation between proplets and core values is established by embedding a given core value into different proplet shells. This is a simple but effective method to enhance the expressive power of the lexicon of a natural language without having to acquire additional core values. For example, the core value *book* may be used as a noun, a verb, or an adj:

6.6.4 EXAMPLES USING *book* IN DIFFERENT PARTS OF SPEECH

Mary loves a good *book* (noun).
Mary *book*ed (verb) a flight to Paris.
Mary is a rather *book*ish (adj) girl.

The lexical *book* proplets used in these contents are defined as follows:

6.6.5 CORE VALUE *book* IN NOUN, VERB, AND ADJ PROPLETS

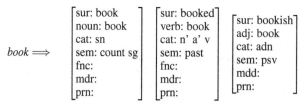

Similar examples are *red* and *square*, which may also be used as the core values of a noun, a verb, and an adj, as in the following contents:

6.6.6 EXAMPLES USING *red* AND *square* IN DIFFERENT PARTS OF SPEECH

Mary preferred the other *red* (noun).
The rising sun *red*dened (verb) the sky.
Mary drank *red* (adj) wine.

Mary's house faces a *square* (noun).
Mary *square*d (verb) her account.
Mary bought a *square* (adj) table.

The lexical methods of using (i) different core (6.6.1) and surface (6.6.2) values in the same proplet shell and (ii) the same core value in different proplet shells (6.6.5) are complemented by (iii) the compositional method of syntax and semantics, resulting in an enormous increase in expressive power. For example, in English the core values *book*, *square*, and *red*, embedded into V, N, and A proplet shells, may be combined as follows:

6.6.7 CORE VALUES IN SYNTACTIC-SEMANTIC COMPOSITION

book$_V$ the red$_A$ square$_N$
book$_V$ the square$_N$ red$_A$
book$_V$ the square$_A$ red$_N$
square$_V$ the red$_A$ book$_N$
square$_V$ the book$_N$ red$_A$
square$_V$ the book$_A$ red$_N$
redden$_V$ the square$_A$ book$_N$
redden$_V$ the book$_N$ square$_A$
redden$_V$ the book$_N$ square$_N$
etc.

As the *literal meaning*$_1$ of complex language signs, each of these examples is welldefined: the core values have straightforward procedural implementations and the syntactic-semantic composition of the proplet shells is grammatically well-formed. For this reason, the purely artificial examples in 6.6.7 make so much sense intuitively when viewed from a grammatical standpoint.

The speaker's *utterance meaning*$_2$, defined as the use of the *literal meaning*$_1$ relative to a context (FoCL'99, 4.3.3), in contrast, seems mostly to fail for the examples in 6.6.7. Which of them can be used literally or nonliterally, or not be used sensibly at all, depends on the context of interpretation, for humans and talking robots alike.

Methodologically, however, we must take into account that linguistic examples as isolated signs do not have a context of interpretation. Therefore, evaluating the utterance meaning$_2$ of the examples 6.6.7 amounts to evaluating how easily they can be supplied intuitively with a virtual context of interpretation. Especially for nonliteral uses, the result depends on the imagination of the cognitive agent doing the evaluation.

For defining the proplets in 6.6.1–6.6.3 and 6.6.5, it is sufficient to represent the core values by means of placeholders like book, red, square, food, eat, apple, or dog. Thereby, the placeholders' literal meaning in English speaking humans is used as a temporary substitute for a procedurally equivalent implementation in a talking robot.

In addition, the use of placeholder core values suffices for the following cognitive procedures of DBS:

6.6.8 COGNITIVE PROCEDURES USING PLACEHOLDER CORE VALUES

1. The time-linear syntactic-semantic *interpretation* in the hear mode (6.4.4),

2. the *storage* of content provided by recognition and inferencing in the Word Bank (Sect. 4.2),

3. the navigation-based semantic-syntactic *production* in the speak mode (Sect. 7.4),

4. the definition of such *meaning relations* as synonymy, antonymy, hypernymy, hyponymy, meronymy, and holonymy as well as cause-effect (Sect. 4.3),

5. the design and implementation of reactor, deductor, and effector *inferences* (Sect. 5.1),

6. the design and implementation of language inferences for adjusting *perspective* (Chaps. 10 and 11), and

7. the *interaction* between the context and language levels (Sect. 6.4).

In other words, computational linguistics can provide a talking robot with a functionally complete cognitive framework by using placeholders to postpone the necessary implementation of artificial recognition and action. The peripheral procedures are needed to supply the placeholder core values with corresponding functional operations.

Without a procedural foundation of the basic core values, the artificial cognition agent will not be able to understand natural language and to act meaningfully. The procedures are used not only (i) to interpret and control the agent's artificial eyes, ears, hands, legs, etc., but also (ii) as the core elements of content, and (iii) as the agent's basic language meanings.

In practice, this means that learning a new word requires the agent not only to remember the correct surface, select the correct proplet shell, and enter the correct (placeholder) core value, but also to acquire the core value's correct procedural implementation. Thereby, complex core value procedures may often be acquired by observation.

Consider, for example, an artificial agent taken to the zoo where it sees for the first time a herd of zebras. By looking at a certain zebra several times and comparing it with the other zebras, the agent automatically derives a new type by distilling the necessary properties within certain ranges, leaving aside the accidental properties of the individuals. Thereby, two kinds of repetition are used: (i) looking several times at the same animal in different postures and (ii) comparing several different instances of the same species (token).[27]

The new zebra type may be constructed in two ways. If the type for horse is already available to the artificial agent, the new type in question may be constructed as a combination of the types for small horse and for black-and-white stripes. If the type for horse is not available, in contrast, the type for zebra may be constructed from more basic types, e.g., for head, body, legs, tail, and black-and-white stripes – such that a later first encounter with a horse will result in a type defined as *big zebra without black-and-white stripes.*

The alternative definitions of a zebra in terms of a horse or a horse in terms of a zebra are equivalent for all practical purposes. Either type will allow the agent to recognize any future zebras, resulting in instantiations (zebra tokens) serving as the (procedural) core values of context proplets. In this way, an artificial agent without language may learn an open number of new concepts.[28] In vision, these concepts are built up as new combinations of elementary, universal types implemented as line, edge, ridge, color, etc. detectors.

Next consider an artificial agent with language learning a new word. After the agent has acquired the concept type for a zebra, a human points at the zebras and utters the French word **zèbre**.[29] Given the similarity between context proplets and language proplets, the artificial agent is able to extend its French vocabulary by (i) copying the context proplet to the language level and (ii) inserting the French surface into the **sur** slot of the copied proplet.

6.6.9 LEARNING A NEW WORD

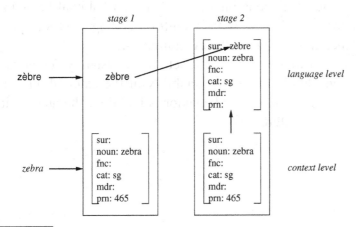

<hr />

[27] In machine learning, this kind of learning is called learning by observation.

[28] In other words, DBS does not attempt to bootstrap all cognition from a small set of universal, basic language concepts, in contradistinction to Wierzbicka (1991).

[29] This surface was chosen to differ from the core value, as in the examples in 6.6.3.

In this way, an artificial agent may learn an open number of words by "contextual definition" – where "contextual" means the (nonlanguage) context of use (and not the language-level co-text).

Modeling the adaptability of cognition in DBS results in a substantial degree of unpredictability in the artificial agent's behavior seen from the outside.[30] For example, (i) the contents in an agent's Word Bank are a rich source of far-reaching personal associations which are inaccessible to an outside observer, (ii) the agent's recent history with its various sources of stress may create short-term sensitivities which are hard to predict, (iii) the more constant leanings of the agent may bring about unexpected preferences, and (iv) the agent may have unusual, individual strategies to realize those preferences.

Corresponding to this unpredictability of an artificial agent seen from the outside is the agent's constant adjustment on the inside to maintain a state of balance. Any parameter values deviating from the "desired" value are recorded by the DBS system and used as triggers to compute compensatory actions. At first, such reactive behavior may be hard to predict from the outside. Over time, however, patterns will emerge in many areas of behavior. They are the result of an agent-internal formation of schemata. A hierarchy of schemata results in levels of abstraction, which each have their own set of inferences.

Individual learning by experience results in more and more predictable behavior. On the one hand, this is necessary as a process of "fitting in." On the other hand, there is the possibility of redundant habits. To interrupt the interminable glide into fruitless routine, a regular, general cleanup of the system may be installed to revise cognition into more productive behavior, based on cost-benefit analysis. The continuous interaction between habitual behavior and innovative behavior is facilitated by their uniform coding as sets of proplets and their time-linear processing with LA-grammars.

Innovative behavior at various levels of abstraction is constantly demanded from the cognitive agent due to the unpredictability of the external and internal environments. In DBS, innovative behavior is based on the automatic derivation and application of inferences.

---

[30] In humans, customs of politeness buffer the inherent uncertainty about another's state of mind. Politeness must be realized also in the artificial agent with language. Given that politeness is ritualized (fixed) behavior, this should not be too difficult.

Part II

The Coding of Content

7. Compositional Semantics

A computational theory of cognition requires (i) completeness of *function*, (ii) completeness of *data coverage*, and (iii) performance in *real time*. (i) is essential for the success of long-term upscaling, (ii) for survival in the agent's ecological niche, and (iii) for ensuring (i) and (ii) in daily performance.

Part I approached functional completeness by modeling the mechanism of natural language communication in the form of a DBS robot with a body, a memory, and external interfaces for recognition and action. Part II turns to completeness of data coverage. The data to be analyzed are kinds of *content*, used in nonlanguage and language cognition alike.

7.1 Forms of Graphical Representation

Programming the cognition of a DBS robot requires the explicit definition of content proplets and pattern proplets. The proplet format is needed for such functional purposes as storage in a Word Bank, activation (retrieval), schema derivation, matching, inferencing, navigation, and language production. However, for a simple, schematic overview of many different kinds of content, an equivalent graphical representation will be more user-friendly and also open the road to the insights and methods of graph theory.

The graphical representation of content in DBS is desigend for a conceptual analysis of the semantic relations of structure.[1] These are functor-argument and coordination, at the elementary, phrasal, and clausal levels (3.5.6). The purpose of the representation is (i) the semantic analysis of different content constructions and (ii) the surface production in different natural languages.

As a brief introduction to the issue at hand, let us compare the graphical representations[2] of the subject-verb relation in (i) Phrase Structure Grammar (PSG) , (ii) Dependency Grammar (DG) , and (iii) Database Semantics (DBS), shown for the sentence **Julia slept**:

---

[1] The semantic relations of structure are distinct from the semantic relations of meaning, such as *synonymy, antonymy, hypernymy, meronymy*, and *holonymy*, as well as *cause and effect*. Cf. Sect. 5.3.

[2] First published in Hausser (2010).

7.1.1 COMPARING REPRESENTATIONS OF THE SUBJECT-VERB RELATION

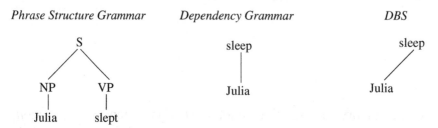

Phrase Structure Grammar   *Dependency Grammar*   *DBS*

The lines (edges) between the nodes (vertices) have completely different interpretations in the three graphs.

In Phrase Structure Grammar (PSG), the lines specify *dominance* and *precedence*. For example, S dominates NP and VP, and NP precedes VP. Thus, the subject-verb relation is not expressed explicitly in the PSG graph, but must be deduced from the part of speech interpretation of the nodes NP and VP, and their configuration in terms of dominance and precedence.[3]

In Dependency Grammar (DG), the line between sleep and Julia specifies Julia as *dependent* on sleep. This does not distinguish between the subject-verb and the object-verb relation, equally analyzed as a dependency.[4]

In Database Semantics (DBS), there are four semantic relations of structure, each represented by its own kind of line: (i) "/" for the subject-verb relation, (ii) "\" for the object-verb relation, (iii) "|" for the modifier-modified relation, and (iv) "−" for the conjunct-conjunct relation (coordination).[5] Furthermore, there are three basic parts of speech, which are represented by the letters N for noun, V for verb, and A for adjective. Given that Julia is an N and sleep is a V, the DBS graph in 7.1.1 explicitly shows the relation between Julia and sleep as N/V, i.e., as subject-verb.

Another example is the modifier-modified relation in the similar but somewhat more complex sentence The little girl slept.

7.1.2 COMPARING DETERMINER-ADJECTIVE-NOUN CONSTRUCTIONS

Phrase Structure Grammar   *Dependency Grammar*   *DBS*

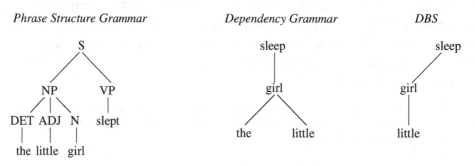

In the PSG graph, the modifier-modified relation is not expressed directly, but must be deduced from the dominance and precedence constellations of the NP, the ADJ, and the N nodes. In the DG graph, the nodes the and little are specified indiscriminately as depending on girl, and girl is specified as depending on sleep. In a DBS graph, however, any "|" line is defined specifically to represent the relation between a modifier and a modified. Given that little is an A and girl is an N, the DBS graph in 7.1.2 characterizes the relation between little and girl explicitly as A|N, i.e., as adnominal (3.5.3, 3.5.4).

7.2 Absorption and Precipitation of Function Words

Another difference between the PSG and DG graphs on the one hand and the DBS graph on the other is in the treatment of function words. For example, the definite article the appears as a node in the PSG and DG graphs of 7.1.2, but not in the DBS graph. This is because the PSG and DG graphs are language analyses, while the DBS graph is an analysis of content. For a representation of content it does not matter whether a certain semantic value such as definiteness is coded by means of a function word, as in English, or by means of a suffix, as in Romanian.[6]

Function words of English are determiners like the, a(n), some, every, all; auxiliary forms like have, had, am, are, were; prepositions like in, on, under, above, below; conjunctions like and, or, that, when; and punctuation signs like comma, full stop, question mark, and exclamation mark.

In accordance with Surface Compositionality, function words are lexically analyzed as proplets which contain certain attributes and/or values representing semantic properties of the word in question. In the hear mode, a function word and its content word are fused and their properties are inherited by the proplet resulting from such a *function word absorption*. For example, the auxiliary was introduces the value past into the resulting verb proplet, the determiner the introduces the value def into the resulting noun proplet, etc. In the speak mode, these same values are used for *function word precipitation*.[7]

---

[3] In contrast, Arc Pair grammar (Johnson and Postal 1980), Relational grammar (Perlmutter 1980), and Functional grammar (Dik 1989, 1997) define grammatical relations such as "subject" and "object" as elementary, as in the DBS approach to the semantic relations of structure.

[4] We are referring here to DG as originally presented by Tesnière (1959). In recent work on DG, such as Liu (2009b) and Hudson (2010), dependencies are differentiated by annotating different roles.

[5] The differentiated interpretation of different graph structures was pioneered by Frege (1879) in his Begriffsschrift.

[6] For example, om (man) is marked as definite with the suffix -ul, as in om-ul (the man).

[7] Function word absorption has the practical advantage that the high frequency words with low semantic significance (Zipf 1932, 1935, 1949) are moved out of the way of a meaningful retrieval.

In function word absorption, content may be absorbed from the content word proplet into the function word proplet or vice versa. For example, in a determiner-noun combination, the relevant properties of the content proplet (noun) are copied into the function proplet (determiner), after which the content proplet (noun) is discarded. Conversely, the relevant properties of the function word proplet (full stop) are copied into the content proplet (verb) in a verb-full stop combination, after which the function word proplet (full stop) is discarded.

The analysis of function words as lexical proplets has to deal with ambiguity, as shown by the following examples:

7.2.1 FUNCTION WORDS WITH DIFFERENT LEXICAL READINGS

1a. Mary has a house *by* the lake.
1b. The book was read *by* Mary.

2a. Mary moved *to* Paris.
2b. Mary tried *to* sleep.

The function words in the (a) variants contribute attributes and values to the proplet resulting from the absorption which are different from those in the (b) variants. This may be handled by providing these function words with more than one lexical analysis (as an exception to Surface Compositionality).

As a consequence of function word absorption, a noun may be phrasal at the surface level, but elementary at the level of content, depending on the number of content words in the surface – and similarly for the other function word constructions. Consider the following examples:

7.2.2 CORRELATING ELEMENTARY/PHRASAL SURFACES AND CONTENTS

elementary surface Julia	phrasal surface the girl	phrasal surface the little girl	
elementary content	elementary content	phrasal content	
$\begin{bmatrix} \text{noun: Julia} \\ \text{cat: nm} \\ \text{sem: sg} \\ \text{fnc: sleep} \\ \text{mdr:} \\ \text{prn: 1} \end{bmatrix}$	$\begin{bmatrix} \text{noun: girl} \\ \text{cat: snp} \\ \text{sem: def sg} \\ \text{fnc: sleep} \\ \text{mdr:} \\ \text{prn: 2} \end{bmatrix}$	$\begin{bmatrix} \text{noun: girl} \\ \text{cat: snp} \\ \text{sem: def sg} \\ \text{fnc: sleep} \\ \text{mdr: little} \\ \text{prn: 3} \end{bmatrix}$	$\begin{bmatrix} \text{noun: little} \\ \text{cat: adn} \\ \text{sem: psv} \\ \text{mdd: girl} \\ \text{nc:} \\ \text{prn: 3} \end{bmatrix}$

While Julia (elementary surface) and the girl (phrasal surface) are each represented by a single proplet (elementary content), the little girl is represented by two proplets, connected by the proposition number 3 and the modifier-modified relation, coded by the mdr and mdd values of the two proplets.

Function word absorption is integrated into the strictly surface compositional, strictly time-linear derivation order of the DBS hear mode. Consider the following example:

7.2.3 HEAR MODE DERIVATION OF The little girl ate an apple.

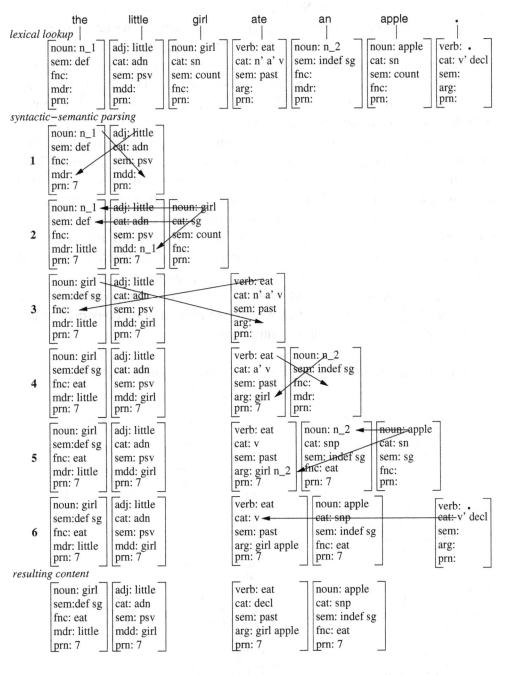

This derivation illustrates the preverbal derivation of the little girl and the postverbal derivation of an apple. The lexical proplets are defined in NLC'06,

Sect. 13.1, while the LA-hear grammar producing the derivation, called LA-hear.2, is defined in NLC'06, Sect. 13.2.

In line 1, the core values of the proplets representing the determiner the and the adnominal little are cross-copied into the respective mdr (modifier) and mdd (modified) slots. In line 2, the core value of the noun proplet *girl* is absorbed into the determiner proplet. This involves (i) replacing the two occurrences of the substitution value n_1, originating from the determiner, with the core value of the *girl* proplet, (ii) copying the cat value sg of the *girl* proplet into the sem slot of the former determiner proplet, and (iii) discarding the *girl* proplet. The gap resulting from the absorption shows up in lines 3–6.

In line 4, the core values of the proplets representing the verb eat and the determiner an are cross-copied into the respective fnc (functor) and arg (argument) slots. In line 5, the *apple* proplet is absorbed into the determiner proplet; this involves (i) replacing the two occurrences of the substitution value n_2 originating from the determiner with the core value of the *apple* proplet, and (ii) discarding the *apple* proplet. The resulting gap appears in line 6.

Another construction with function word absorption is a prepositional phrase like in the garden, compared below with the phrasal noun construction the garden, both as used in a sentence-initial position:

7.2.4 COMPARING DIFFERENT FUNCTION WORD ABSORPTIONS

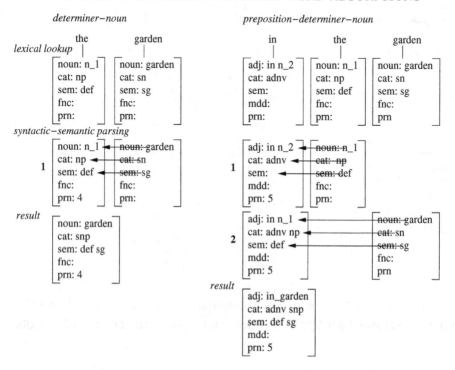

On the left, the lexical determiner proplet representing the surface **the** has the core attribute **noun** because after absorbing the noun proplet for **garden** the resulting proplet will be a noun. The single step of the determiner-noun derivation substitutes the **n_1** value of the determiner proplet with the core value of the *garden* proplet, adds the **sg** value of the *garden* proplet to the **sem** attribute of the determiner proplet, and discards the *garden* proplet.

On the right, the lexical preposition proplet representing the surface **in** has the core value **adj** because, after absorbing the determiner proplet for **the** and the noun proplet for **garden**, the resulting proplet is a phrasal adjective.[8] Step 1 of the time-linear preposition-determiner-noun derivation combines the two lexical function word proplets for **in** and **the** into one **adj** proplet. Thereby the substitution value **n_2** in the preposition proplet is replaced with the substitution value **n_1** of the determiner proplet, the **def** value of the determiner proplet is added to the **sem** attribute of the preposition proplet, and the determiner proplet is discarded.

Step 2 combines the **adj** proplet resulting from step 1 with the lexical *garden* proplet into an **adj** proplet. This is based on replacing the **n_1** substitution value in the **adj** proplet with the core value of the *garden* proplet, adding the **sg** value of the *garden* proplet to the **sem** attribute of the **adj** proplet, and discarding the *garden* proplet.

Thus, a prepositional phrase, as in **Julia slept in the garden.**, and an elementary adjective, as in **Julia slept there.**, may be represented by the same single node, A.

7.2.5 PREPOSITIONAL PHRASE AS ELEMENTARY ADJECTIVE

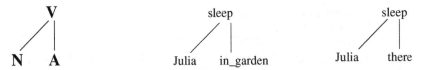

The modifier-modified relation, indicated by "|," connects an A and a V, and constitutes the adverbial[9] use of an adjective. The crucial difference between the two elementary content adverbials is the sign kind[10] of their core values: the core value of **in_garden** is a *symbol*, while that of **there** is an *indexical*.

---

[8] As shown in 3.5.3, DBS uses the **cat** value **adnv** for modifiers which may be applied to verbs or nouns, e.g., **fast**, the **cat** value **adn** (adnominal) for modifiers restricted morphologically to nouns, e.g., **beautiful**, and the **cat** value **adv** (adverbial) for modifiers restricted morphologically to verbs, e.g., **beautifully**. Because prepositional phrases may be used adnominally as well as adverbially they have the **cat** value **adnv**.

[9] This is in contradistinction to the adnominal use illustrated in 7.1.2. See NLC'06, Chap. 15, for a detailed analysis of elementary and phrasal adjectives in adnominal and adverbial use.

[10] Cf. FoCL'99, Chap. 6.

7.3 Deriving DBS Graphs from Sets of Proplets

The leftmost graph in 7.2.5 is called a part of speech signature, or *signature* for short, while the other graphs are called semantic relations graphs, or *SRGs* for short.[11] A signature resembles a phrase structure tree insofar as it uses letters for nodes, while an SRG resembles a DG graph insofar as it uses English words instead (7.1.1, 7.1.2). SRGs and signatures are conceptual superstructures on top of sets of proplets serving as the full-fledged representation of content.

A signature is more general than an SRG because it abstracts away from the content words by replacing them with letters representing their parts of speech.[12] An SRG, in contrast, is more intuitive in that it illustrates a content construction concretely with actual concepts represented by English words. For the software procedures of DBS, however, the more differentiated proplet representation of content must be used. This raises the following questions:

7.3.1 ON RELATING PROPLET SETS TO DBS GRAPHS

- What is the nature of the relation between the proplet representation of a content and the corresponding SRG or signature?
- Can an SRG or a signature be derived automatically from a set of proplets representing a certain content?

Consider the content derived in 7.2.3 as an order-free set of proplets:

7.3.2 CONTENT CORRESPONDING TO The little girl ate an apple.

$$
\begin{bmatrix} \text{noun: girl} \\ \text{sem: def sg} \\ \text{fnc: eat} \\ \text{mdr: little} \\ \text{prn: 7} \end{bmatrix}
\begin{bmatrix} \text{adj: little} \\ \text{cat: adn} \\ \text{sem: psv} \\ \text{mdd: girl} \\ \text{prn: 7} \end{bmatrix}
\begin{bmatrix} \text{verb: eat} \\ \text{cat: decl} \\ \text{sem: past} \\ \text{arg: girl apple} \\ \text{prn: 7} \end{bmatrix}
\begin{bmatrix} \text{noun: apple} \\ \text{cat: snp} \\ \text{sem: indef sg} \\ \text{fnc: eat} \\ \text{prn: 7} \end{bmatrix}
$$

The first step of turning this set of proplets automatically into an SRG or a signature is to form all possible unordered pairs of related proplets, i.e., {*girl, eat*}, {*apple, eat*}, and {*little, girl*}, and to apply the following schemata to each pair:

7.3.3 SCHEMATA INTERPRETING TRANSPARENT INTRAPROP. RELATIONS

(1) noun/verb
$$\begin{bmatrix} \text{noun: } \alpha \\ \text{fnc: } \beta \end{bmatrix} \begin{bmatrix} \text{verb: } \beta \\ \text{arg: } \alpha \text{ X} \end{bmatrix}$$

(2) noun\verb
$$\begin{bmatrix} \text{noun: } \alpha \\ \text{fnc: } \beta \end{bmatrix} \begin{bmatrix} \text{verb: } \beta \\ \text{arg: } \gamma \text{ X } \alpha \end{bmatrix}$$

(3) adjective|noun
$$\begin{bmatrix} \text{adj: } \alpha \\ \text{mdd: } \beta \end{bmatrix} \begin{bmatrix} \text{noun: } \beta \\ \text{mdr: X } \alpha \text{ Y} \end{bmatrix}$$

(4) adjective|verb
$$\begin{bmatrix} \text{adj: } \alpha \\ \text{mdd: } \beta \end{bmatrix} \begin{bmatrix} \text{verb: } \beta \\ \text{mdr: X } \alpha \text{ Y} \end{bmatrix}$$

(5) noun− noun
$$\begin{bmatrix} \text{noun: } \alpha \\ \text{nc: } \beta \end{bmatrix} \begin{bmatrix} \text{noun: } \beta \\ \text{pc: } \alpha \end{bmatrix}$$

(6) verb−verb
$$\begin{bmatrix} \text{verb: } \alpha \\ \text{nc: } \beta \end{bmatrix} \begin{bmatrix} \text{verb: } \beta \\ \text{pc: } \alpha \end{bmatrix}$$

(7) adjective−adjective
$$\begin{bmatrix} \text{adj: } \alpha \\ \text{nc: } \beta \end{bmatrix} \begin{bmatrix} \text{adj: } \beta \\ \text{pc: } \alpha \end{bmatrix}$$

The only pair matching schema (1) is {*girl, eat*}, the only pair matching schema (2) is {*apple, eat*}, and the only pair matching schema (3) is {*little, girl*}.

Arranging and connecting the proplets in accordance with the successful matches between 7.3.2 and 7.3.3 results in the following graph:

7.3.4 DBS GRAPH BASED ON PROPLETS (PROPLET GRAPH)

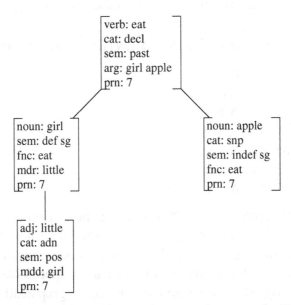

In a second step, this proplet graph may be turned automatically into an SRG or a signature. The SRG is obtained by replacing each proplet with its core *value*, while the signature is obtained by replacing each proplet with a letter representing the part of speech of its core *attribute*:

7.3.5 RESULTING SRG AND SIGNATURE

(i) semantic relations graph (SRG)   *(ii) signature*

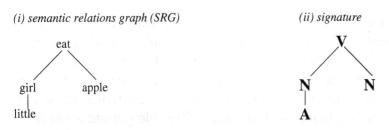

---

[11] As pointed out by Haitao Liu, an SRG resembles the **stemma réelle** while a signature resembles the **stemma virtuelle** of Tesnière (1959). See also Ágel et al. (eds.) (2006), p. 1316.

[12] Structure A is homomorph to a structure B iff (i) each elementary node in A has a counterpart in B and (ii) each relation in A has a corresponding relation in B. Cf. FoCL'99, Sect. 21.3.

The SRG and the signature in 7.3.5 are each *homomorph* to the proplet graph
7.3.4. The direction of the homomorphism from SRGs and signatures to proplet graphs is necessitated by the more differentiated structure of proplets as
compared to the corresponding nodes in SRGs or signatures.

A pair of directly connected nodes in a signature constitutes an *elementary
signature*. In running text, elementary signatures are written as two nodes with
a line in between, e.g., N/V. There are the following seven elementary signatures in DBS which are (i) intrapropositional and (ii) transparent:[13]

7.3.6 THE SEVEN TRANSPARENT SEMANTIC RELATIONS OF STRUCTURE

subject-verb:	1. N/V
object-verb:	2. N\V
adjective-noun:	3. A|N
adjective-verb:	4. A|V
conjunct-conjunct:	5. N−N
	6. V−V
	7. A−A

For example, the signature in 7.3.5 is based on the elementary signatures N/V
(subject-verb), N\V (object-verb), and A|N (adnominal-noun).

It holds in general that the order of the letters in an elementary signature like
N/V does not indicate any primacy of one part of speech over the other. For
now (cf. Sect. 7.6), let us motivate the letter order graphically: in the case of
elementary "/," " \," and "|" signatures, the first node is shown lower than the
second node in the graph (7.3.4), while in elementary "−" signatures the first
node is shown to the left of the second node in the graph (7.4.4).

7.4 Producing Natural Language Surfaces from Content

For natural language production from content, the (i) SRG and the (ii) signature are complemented by two additional structures, called the (iii) *numbered
arcs graph* or NAG[14] and the (iv) *surface realization*. Together, these conceptual structures provide four simultaneous views (in the conceptual sense) on
a given content. This is illustrated next with the graphical representation of
the content corresponding to the English surface **The little girl ate an apple.**,
familiar from 7.2.3, 7.3.2, 7.3.4 and 7.3.5:

---

[13] For the inverse constructions, i.e., (i) extrapropositional and (ii) opaque signatures, see Sect. 7.5.
[14] For graph-theoretical and linguistic constraints on NAGs, see Sect. 9.1.

7.4.1 THE FOUR DBS VIEWS ON A CONTENT AND ITS SURFACE

(i) *semantic relations graph (SRG)*   (iii) *numbered arcs graph (NAG)*

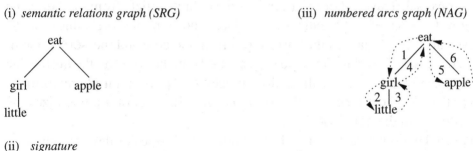

(ii) *signature*

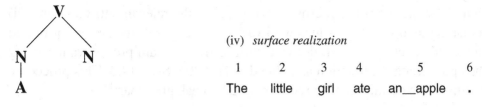

(iv) *surface realization*

1	2	3	4	5	6
The	little	girl	ate	an_apple	.

The SRG (i) and the signature (ii) are the same as in 7.3.4. The NAG (iii) shows the time-linear activation order of the SRG. The surface realization (iv) shows the use of the traversal order for production in English.

A numbered arcs graph may be defined for an SRG, a signature, or a proplet graph. Consider the NAG based on the proplet graph 7.3.4:

7.4.2 NUMBERED ARCS GRAPH BASED ON PROPLETS (PROPLET NAG)

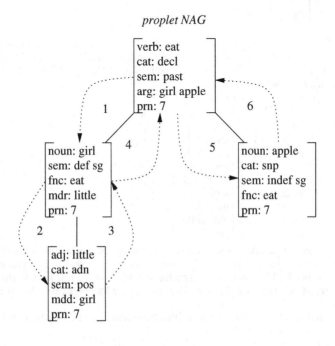

proplet NAG

The word form surfaces are always realized from the *goal* proplet (goal node) of a traversal. Traversal of arc 1 realizes the definite article from the **def** value in *girl*, traversal of arc 2 realizes the surface **little** from the core value of *little*, traversal of arc 3 realizes the surface **girl** from the core and the **sem** value of *girl*, traversal of arc 4 realizes the surface **ate** from the core and the **sem** value of *eat*, traversal of arc 5 realizes the surfaces **an apple** (multiple realization) from the **sem** and the core value of *apple*, and traversal of arc 6 realizes the full stop from the **cat** value of *eat*.

Given the proplet NAG 7.4.2, the definition of LA-speak rules for English is easy: 1. take the proplet pair (elementary signature) of the traversal in question, 2. define the rule patterns by (i) selecting the relevant attributes and (ii) replacing their constant values with variables, 3. specify the successor rule(s) in the rule package, and 4. derive a surface from certain proplet values using the appropriate lex function as defined in NLC'06, Sect. 14.3. This procedure results in the definition of the following LA-speak grammar:[15]

7.4.3 LA-SPEAK GRAMMAR FOR THE PROPLET NAG 7.4.2

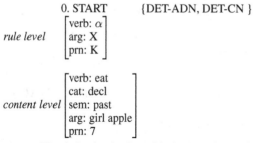

0. START   {DET-ADN, DET-CN }

rule level
$$\begin{bmatrix} \text{verb: } \alpha \\ \text{arg: X} \\ \text{prn: K} \end{bmatrix}$$

content level
$$\begin{bmatrix} \text{verb: eat} \\ \text{cat: decl} \\ \text{sem: past} \\ \text{arg: girl apple} \\ \text{prn: 7} \end{bmatrix}$$

Comment: The derivation begins with the Start State of the grammar, which matches the verb proplet *eat* and activates the Start rule package containing the rules DET-ADN and DET-CN.

Next, arc 1 is traversed by the following application of the rule DET-ADN (for determiner to be followed by an adnominal):

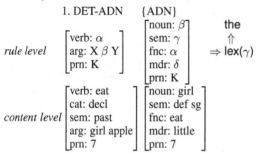

1. DET-ADN   {ADN}

rule level
$$\begin{bmatrix} \text{verb: } \alpha \\ \text{arg: X } \beta \text{ Y} \\ \text{prn: K} \end{bmatrix} \begin{bmatrix} \text{noun: } \beta \\ \text{sem: } \gamma \\ \text{fnc: } \alpha \\ \text{mdr: } \delta \\ \text{prn: K} \end{bmatrix}$$
the
⇑
⇒ lex(γ)

content level
$$\begin{bmatrix} \text{verb: eat} \\ \text{cat: decl} \\ \text{sem: past} \\ \text{arg: girl apple} \\ \text{prn: 7} \end{bmatrix} \begin{bmatrix} \text{noun: girl} \\ \text{sem: def sg} \\ \text{fnc: eat} \\ \text{mdr: little} \\ \text{prn: 7} \end{bmatrix}$$

Comment: The proplet patterns of the rule level match corresponding proplets in the NAG 7.4.2. The rule package activates ADN as the next rule. The output arrow ⇒ shows the application of the lex function defined in NLC'06, 14.3.2. Based on binding the variable γ of the rule level to the **sem** values **def sg** of the goal proplet *girl*, the lex function realizes the surface **the**. Next, arc 2 is traversed.

---

[15] The definition of LA-speak in 7.4.3 benefitted from comments and suggestions by J. Handl.

2. ADN   {CN}

$$\text{rule level} \quad \begin{bmatrix} \text{noun: } \beta \\ \text{mdr: } X\,\alpha\,Y \\ \text{prn: } K \end{bmatrix} \begin{bmatrix} \text{adj: } \alpha \\ \text{cat: } \gamma \\ \text{mdd: } \beta \\ \text{prn: } K \end{bmatrix} \quad \begin{matrix} \text{little} \\ \Uparrow \\ \Rightarrow \text{lex}(\alpha\,\gamma) \end{matrix}$$

$$\text{content level} \quad \begin{bmatrix} \text{noun: girl} \\ \text{sem: def sg} \\ \text{fnc: eat} \\ \text{mdr: little} \\ \text{prn: 7} \end{bmatrix} \begin{bmatrix} \text{adj: little} \\ \text{cat: adn} \\ \text{sem: psv} \\ \text{mdd: girl} \\ \text{prn: 7} \end{bmatrix}$$

Comment: After matching and binding the variable α of the rule level to the mdd value little of the goal proplet, a lex function refined from NLC'06, 14.3.7, realizes the corresponding English surface.

Next, arc 3 in 7.4.2 is traversed by the following application of the rule CN (for common noun):

3. CN   {FV, PNC}

$$\text{rule level} \quad \begin{bmatrix} \text{adj: } \alpha \\ \text{mdd: } \beta \\ \text{prn: } K \end{bmatrix} \begin{bmatrix} \text{noun: } \beta \\ \text{cat: } \gamma \\ \text{mdr: } \alpha \\ \text{prn: } K \end{bmatrix} \quad \begin{matrix} \text{girl} \\ \Uparrow \\ \Rightarrow \text{lex}(\beta\,\gamma) \end{matrix}$$

$$\text{content level} \quad \begin{bmatrix} \text{adj: little} \\ \text{cat: adn} \\ \text{sem: psv} \\ \text{mdd: girl} \\ \text{prn: 7} \end{bmatrix} \begin{bmatrix} \text{noun: girl} \\ \text{cat: sn} \\ \text{fnc: eat} \\ \text{mdr: little} \\ \text{prn: 7} \end{bmatrix}$$

Comment: After matching and binding the variable β to the noun value girl, a lex function refined from NLC'06, 14.3.3, realizes the English surface. A successful rule application activates FV and PNC as the next rules.

Next, arc 4 is traversed by the following application of the rule FV (for finite verb):

4. FV   {DET-ADJ, DET-CN}

$$\text{rule level} \quad \begin{bmatrix} \text{noun: } \beta \\ \text{fnc: } \alpha \\ \text{prn: } K \end{bmatrix} \begin{bmatrix} \text{verb: } \alpha \\ \text{arg: } \beta\,Y \\ \text{sem: } \gamma \\ \text{prn: } K \end{bmatrix} \quad \begin{matrix} \text{ate} \\ \Uparrow \\ \Rightarrow \text{lex}(\alpha\,\gamma) \end{matrix}$$

$$\text{content level} \quad \begin{bmatrix} \text{noun: girl} \\ \text{sem: def sg} \\ \text{fnc: eat} \\ \text{mdr: little} \\ \text{prn: 7} \end{bmatrix} \begin{bmatrix} \text{verb: eat} \\ \text{cat: decl} \\ \text{sem: past} \\ \text{arg: girl apple} \\ \text{prn: 7} \end{bmatrix}$$

Comment: After matching and binding the variables α and γ of the rule level to the verb value eat and the cat value past, respectively, of the goal proplet, the lex function defined in NLC'06, 14.3.4, realizes the English surface ate. A successful rule application activates DET-ADN and DET-CN.

Next, arc 5 is traversed by the application of the rule DET-CN (for determiner+common noun):

5. DET-CN   {FV, PNC}

$$\text{rule level} \quad \begin{bmatrix} \text{verb: } \alpha \\ \text{arg: } X\,\beta\,Y \\ \text{prn: } K \end{bmatrix} \begin{bmatrix} \text{noun: } \beta \\ \text{sem: } \gamma \\ \text{fnc: } \alpha \\ \text{mdr: NIL} \\ \text{prn: } K \end{bmatrix} \quad \begin{matrix} \text{an apple} \\ \Uparrow \\ \Rightarrow \text{lex}(\gamma\,\beta) \end{matrix}$$

$$\text{content level} \quad \begin{bmatrix} \text{verb: eat} \\ \text{cat: decl} \\ \text{sem: past} \\ \text{arg: girl apple} \\ \text{prn: 7} \end{bmatrix} \begin{bmatrix} \text{noun: apple} \\ \text{sem: indef sg} \\ \text{fnc: eat} \\ \text{mdr:} \\ \text{prn: 7} \end{bmatrix}$$

Comment: Rule DET-CN applies the lex functions defined in NLC'06, 14.3.2 and 14.3.3, to realize **an apple** (double realization), using the **sem** values **indef sg** of *apple* for the determiner and the **noun** value for the noun. A successful rule application activates FV and PNC as the next rules.

Finally, arc 6 is traversed using the following rule, PNC (for punctuation):

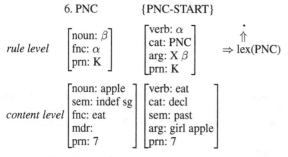

$$
\begin{array}{lll}
\textit{rule level} &
\begin{bmatrix} \text{noun: } \beta \\ \text{fnc: } \alpha \\ \text{prn: K} \end{bmatrix}
&
\begin{bmatrix} \text{verb: } \alpha \\ \text{cat: PNC} \\ \text{arg: X } \beta \\ \text{prn: K} \end{bmatrix}
&
\begin{array}{c} \bullet \\ \Uparrow \\ \Rightarrow \text{lex(PNC)} \end{array}
\end{array}
$$

$$
\begin{array}{ll}
\textit{content level} &
\begin{bmatrix} \text{noun: apple} \\ \text{sem: indef sg} \\ \text{fnc: eat} \\ \text{mdr:} \\ \text{prn: 7} \end{bmatrix}
&
\begin{bmatrix} \text{verb: eat} \\ \text{cat: decl} \\ \text{sem: past} \\ \text{arg: girl apple} \\ \text{prn: 7} \end{bmatrix}
\end{array}
$$

Comment: The rule PNC applies the lex function defined in NLC'06, 14.3.8, to realize the full stop, using the **cat** value **decl** of the *eat* proplet. A successful application activates PNC+START as the next rule (for a possible traversal to a next proposition) and the Final State ST_F, defined as $\{([\text{verb}: \alpha] \, RP_{PCN})\}$.

The translation of LA-speak rules, specified in the manner shown in 7.4.3, into the JSLIM software is routine.[16]

The conceptual analysis underlying such a proplet-based software may begin with the DBS graph structures, because they are simpler than the corresponding LA-speak grammar. Consider, for example, extrapropositional coordination. The content of **Julia sang. Sue slept. John read.**, shown in 3.2.5 as a set of proplets, may be analyzed as the following graph structure:

7.4.4 REPRESENTING AN EXTRAPROPOSITIONAL COORDINATION

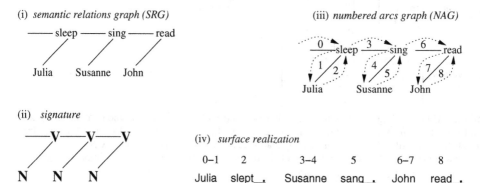

In text- or dialogue-initial position, the initial "−" coordination line without a left-hand node serves as the *start line*. Subsequent extrapropositional "−" lines have a left-hand node, representing the verb of the preceding proposition.[17]

---

[16] While NLC'06 approached natural language production by defining an LA-think grammar first and then adding the LA-speak grammar for English on top of it, our present approach is to define the LA-speak grammar first and to obtain the LA-think grammar by omitting the part with the ⇒ and ⇑ arrows in the LA-speak rules, i.e., the surface output based on the lex functions.

The surface realization of 7.4.4 shows not only that a single navigation step can realize more than one surface (cf. steps 2, 5, 8), but that a single surface may require more than one navigation step (cf. steps 0–1, 3–4, and 6–7).

The possibility of empty traversals[18] in combination with traversals which realize more than one surface[19] allows for producing more than one surface order from a given numbered arcs graph. As an example, consider the six possible permutations of {dog, find, bone}, as needed for language production from a given content in a free word order language like Russian:

7.4.5 RUSSIAN WORD ORDER BASED ON ALTERNATIVE TRAVERSALS

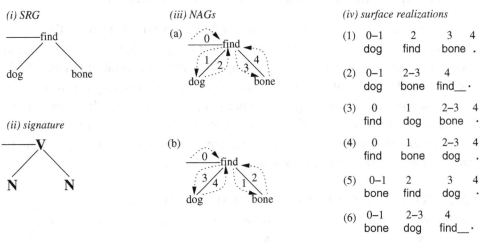

(i) SRG	(iii) NAGs	(iv) surface realizations

There are (i) one SRG, (ii) one signature, (iii) two NAGs (a) and (b), and (iv) six surface realizations (1)–(6). The NAGs begin with arc 0. Given that surfaces may be realized only from the goal proplet of a navigation step, arc 0 is not only motivated by extrapropositional coordination (7.4.4), but is also needed for verb-first surfaces, such as find dog bone.

Let us go through the surface realizations step-by-step. Realizations (1) and (2) are based on NAG (a); they share a 0–1 traversal to realize dog. (1) continues with an arc 2 traversal to produce find, an arc 3 traversal to produce bone, and an arc 4 traversal to produce the full stop. (2) uses an arcs 2–3 traversal to produce bone and an arc 4 traversal to produce find_.

Surface realizations (3) and (4) share an arc 0 traversal to realize find. (3) is based on NAG (a) and continues with an arc 1 traversal to produce dog, a 2–3 traversal to produce bone, and an arc 4 traversal to produce the full

---

[17] Cf. NLC'06, Chaps. 11 and 12, for a detailed analysis of extrapropositional coordination. The combination of an intra- and an extrapropositional verb coordination is shown in NLC'06, 9.1.2.

[18] I.e., traversals which do not realize a surface, like those of arcs 0, 3, and 6 in 7.4.4.

[19] Like those of arcs 2, 5, and 8 in 7.4.4.

stop. Surface realization (4) is based on NAG (b) and continues with an arc 1 traversal to produce **bone**, a 2–3 traversal to produce **dog**, and an arc 4 traversal to produce the full stop.

Surface realizations (5) and (6) are based on NAG (b). They share a 0–1 traversal to realize **bone**. (5) continues with an arc 2 traversal to produce **find**, an arc 3 traversal to produce **dog**, and an arc 4 traversal to produce the full stop. (6) uses an arcs 2–3 traversal to produce **dog** and an arc 4 traversal to produce **find_**.

7.5 Transparent vs. Opaque Functor-Argument

The elementary signatures in 7.3.6 are called transparent because the corresponding combination of proplets uses their original lexical form, without having to add new attributes for coding the semantic relation in question. The treatment of subclauses in NLC'06, Chap. 7, however, requires also $V/_xV$ (subject sentence), $V\backslash_xV$ (object sentence), $V|_xN$ (sentential adnominal modifier, a.k.a. relative clause), and $V|_xV$ (sentential adverbial modifier) relations.[20] Consider the following signatures of extrapropositional functor-arguments in comparison with their intrapropositional counterparts:[21]

7.5.1 EXTRA- VS. INTRAPROPOSITIONAL FA STRUCTURES

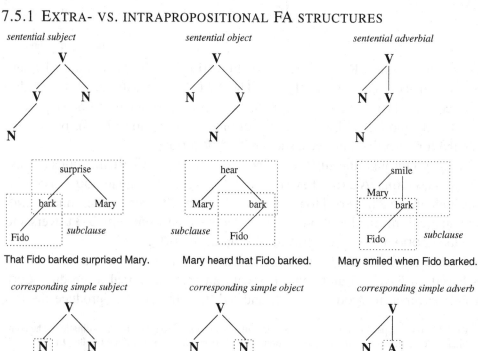

The double function of **bark** as (i) the verb of the subclause and (ii) the noun or adverbial of the main clause is shown by the dotted rectangles in their SRGs. No such double function exists in the corresponding elementary nouns and adverbials, as shown by the signatures at the bottom.

Extrapropositional functor-argument relations are opaque[22] because the relation between the lower verb and the higher verb (or noun in relative clauses; cf. Sects. 9.3 and 11.5) requires attributes and values which differ in kind from those of the corresponding transparent intrapropositional relations listed in 7.3.6. More specifically, the lower verb requires an additional attribute (either **fnc** or **mdd**) and (ii) the higher verb or higher noun has a verbal value in its **arg** or **mdr** slot – attributes which are normally reserved for nominal or adjectival values, respectively.

As a concrete example, consider the proplet representation of the sentential subject construction in 7.5.1:[23]

7.5.2 PROPLET REPRESENTATION OF **That Fido barked surprised Mary.**

$$
\begin{bmatrix} \text{noun: Fido} \\ \text{cat: nm} \\ \text{sem: sg} \\ \text{fnc: bark} \\ \text{mdr:} \\ \text{prn: 27} \end{bmatrix}
\begin{bmatrix} \text{verb: bark} \\ \text{cat: v} \\ \text{sem: past} \\ \text{arg: Fido} \\ \text{fnc: (surprise 28)} \\ \text{prn: 27} \end{bmatrix}
\begin{bmatrix} \text{verb: surprise} \\ \text{cat: decl} \\ \text{sem: past} \\ \text{arg: (bark 27) Mary} \\ \text{mdr:} \\ \text{prn: 28} \end{bmatrix}
\begin{bmatrix} \text{noun: Mary} \\ \text{cat: nm} \\ \text{sem: sg} \\ \text{fnc: surprise} \\ \text{mdr:} \\ \text{prn: 28} \end{bmatrix}
$$

The *bark* proplet has the core attribute **verb**, but unlike a regular verb it also has the attribute **fnc**, which is normally reserved for nouns. Its value (**surprise 28**) establishes the extrapropositional subject/$_x$verb relation between the subject sentence and the verb **surprise** of the main clause. Conversely, *surprise* has (**bark 27**) in the subject position of its **arg** slot.[24]

The semantic relations between the verb of the lower sentence and the verb of the higher sentence are strictly local. For example, in 7.5.2, the remainder of the subject sentence comes into play only insofar as *Fido* is connected to, and is dragged along by, *bark*. The remainder of the main clause comes into play only insofar as *Mary* is connected to, and is dragged along by, *surprise*.

---

[20] In analyses not concerned with extrapropositional coordination the arc 0 is optional.

[21] The subscripts x in the elementary signatures indicate an extrapropositional relation.

[22] The *transparent/opaque* distinction was used by Quine (1960) for another purpose, namely to distinguish between contexts called *even/uneven* by Frege (1892) and *extensional/intensional* by Montague (1974); cf. FoCL'99, Chap. 20.

[23] Cf. NLC'06, 7.2.1, for an analogous hear mode derivation.

[24] The proplets of the subclause have the **prn** value 27, while those of the main clause have the **prn** value 28. Because of these different **prn** numbers, the value of the **fnc** attribute of *bark* is specified as the address (**surprise 28**) and the first value of the **arg** attribute of *surprise* is specified as the address (**bark 27**). See Sect. 4.4 for a more detailed discussion of symbolic addresses and pointers.

This is different from structuralism and its later ramifications, which are based on possible *substitutions* and combine whole constituents. The conglomeration into constituents is shown by the nonterminal nodes in PSG trees (e.g., 7.1.1 and 7.1.2). In DBS graphs, nonterminals are absolutely absent.

In addition to the opaque extrapropositonal FA (functor-argument) relations, there are also opaque intrapropositional FA relations in natural language. As an example, consider the prepositional object in Julia put the flowers in a vase., which illustrates an opaque A\V object-verb relation: [25]

7.5.3 THREE-PLACE VERB WITH PREPOSITIONAL PHRASE AS OBJECT
Julia put the flowers in a vase.

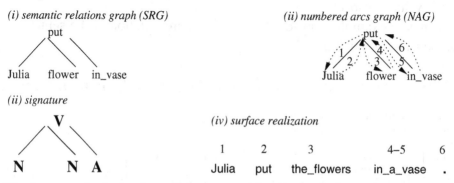

(i) semantic relations graph (SRG)

(ii) numbered arcs graph (NAG)

(ii) signature

(iv) surface realization

	1	2	3	4–5	6
	Julia	put	the_flowers	in_a_vase	.

The hear mode derivation of the prepositional phrase in a vase is analogous to that in 7.2.4 and results in a single proplet, represented in the signature as the A node. The content has the following proplet representation:

7.5.4 OPAQUE INTRAPROPOSITIONAL OBJECT, CONTENT OF 7.5.3

$$
\begin{bmatrix} \text{noun: Julia} \\ \text{cat: nm} \\ \text{sem: sg} \\ \text{fnc: put} \\ \text{mdr:} \\ \text{prn: 26} \end{bmatrix}
\begin{bmatrix} \text{verb: put} \\ \text{cat: decl} \\ \text{sem: past} \\ \text{arg: Julia flower in\_vase} \\ \text{mdr:} \\ \text{prn: 26} \end{bmatrix}
\begin{bmatrix} \text{noun: flower} \\ \text{cat: sn} \\ \text{sem: def pl} \\ \text{fnc: put} \\ \text{mdr:} \\ \text{prn: 26} \end{bmatrix}
\begin{bmatrix} \text{adj: in\_vase} \\ \text{cat: adnv snp} \\ \text{sem: indef sg} \\ \text{fnc: put} \\ \text{mdr:} \\ \text{prn: 26} \end{bmatrix}
$$

The construction is opaque because the adj proplet *in_vase* has an additional fnc attribute, and the verb proplet *put* has an adj argument instead of a noun.

As a lexical proplet, *vase* is an N, yet at the content level the corresponding single proplet representing in a vase is an A. Thus the relation between the *put* proplet and the prepositional object is N\V (transparent) based on the lexical proplet of *vase*, but is A\V (opaque) based on the content proplet *in_vase*.

---

[25] Opaque constructions are frequent in Chinese, for example.

Conversely, the subject-predicate relation in the English copula construction **Julia is a doctor** is N/N (opaque) based on the lexical proplets *Julia* and *doctor*, but is N/V (transparent) based on the content proplets. This is shown by the following hear mode derivation:

7.5.5 HEAR MODE DERIVATION OF **Julia is a doctor**.

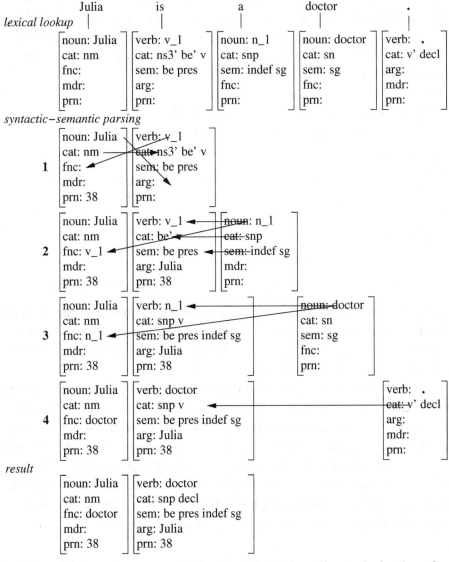

In this strictly surface compositional, strictly time-linear derivation, five lexical proplets are reduced to two content proplets based on function word absorption. In line 1, the core value **Julia** and the substitution value **v_1** are cross-copied into **fnc** and **arg** slots, and the **ns3'** valency position in the **cat**

slot of the *is* proplet is canceled. In line 2, the two occurrences of the substitution value **v_1** are replaced by another substitution value, **n_1**, provided by the determiner. Also, the values **indef sg** are added to the **sem** slot of the *is* proplet. In line 3, the two occurrences of the substitution value **n_1** are replaced by the core value of the *doctor* proplet, which is then discarded. The result is a *doctor* proplet with the core attribute **verb**.

The *doctor* proplet is opaque insofar as the core attribute **verb** has the nominal value **doctor** and the **sem** attribute mixes the nominal values **indef sg** with the verbal value **pres**. The *Julia* proplet is opaque in that its **fnc** attribute has the nominal value **doctor**. The signature, however, shows an apparently transparent N/V structure:

7.5.6 DBS GRAPH ANALYSIS OF Julia is a doctor.

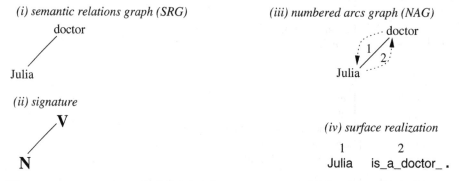

(i) semantic relations graph (SRG)

doctor

Julia

(ii) signature

V

N

(iii) numbered arcs graph (NAG)

doctor

1
2

Julia

(iv) surface realization

1   2
Julia   is_a_doctor_ .

Thus, a content construction may be opaque even if the signature appears to be transparent. Traversal of arc 1 realizes **Julia**, while traversal of arc 2 realizes **is_a_doctor_.**, utilizing the values **be**, **indef**, **sg**, **decl**, and **doctor** of the *doctor* proplet (see result of 7.5.5).

Similarly, the subject-predicate relation in the English copula construction **Fido is hungry** is N/A (opaque) based on the lexical proplets, but N/V (transparent) based on the content proplets. Here the change from the lexical core attribute A to the content core attribute V likewise originates in the function word **is** into which the lexical core value, i.e., **hungry**, is absorbed. Because the DBS representation of semantic relations of structure applies to the level of content, the PoS participating in these semantic relations are determined by the core attributes of the content (and not the lexical) proplets.

7.6 Possible and Actual Semantic Relations of Structure

Intrapropositionally, the number of elementary signatures consisting of the nodes N, V, and A and the lines "/," "\," "|," and "−" is 36. How many of

these are actually used in English? As an initial, tentative approach, consider
the following three lists. The first shows the 12 possible elementary signatures
(binary semantic relations of structure) beginning with N:

7.6.1 TRANSPARENT AND OPAQUE RELATIONS BEGINNING WITH N

1 N/N ?
2 N/V subject/verb (transparent)
3 N/A ?

4 N\N ?
5 N\V object\verb (transparent)
6 N\A ?

7 N|N ?
8 N|V ?
9 N|A ?

10 N−N noun–noun conjunction (transparent)
11 N−V ?
12 N−A ?

Nine of the elementary signatures do not appear to be used in English (indicated by ?), three are transparent (2, 5, 10), and none is opaque.

Next consider the list of elementary signatures beginning with V:

7.6.2 TRANSPARENT AND OPAQUE RELATIONS BEGINNING WITH V

1 V/N ?
2 V/V infinitive_subject/verb (opaque)
3 V/A ?

4 V\N ?
5 V\V infinitive_object\verb (opaque)
6 V\A ?

7 V|N progressive_verb|noun, infinitive|noun (opaque)
8 V|V ?
9 V|A ?

10 V−N ?
11 V−V verb–verb (transparent)
12 V−A ?

One elementary signature is transparent (11), three are opaque (2, 5, 7), and
the remaining eight do not appear to be used in English.

Finally consider the list of elementary signatures beginning with A:

7.6.3 TRANSPARENT AND OPAQUE RELATIONS BEGINNING WITH A

1 A/N ?
2 A/V ?
3 A/A ?

4 A\N ?
5 A\V prepositional_object\verb (opaque)
6 A\A ?

7 A|N adj|noun (transparent)
8 A|V adj|verb (transparent)
9 A|A ?

10 A−N ?
11 A−V ?
12 A−A adj-adj coordination (transparent)

Three elementary signatures are transparent (7, 8, 12), one is opaque (5), and the remaining eight do not appear to be used in English.

In summary, of the 36 possible intrapropositional elementary signatures, English seems to use only 11, of which seven are transparent and four are opaque:

7.6.4 TRANSPARENT VS. OPAQUE INTRAPROPOSITIONAL RELATIONS

transparent intrapropositional relations
1. N/V (subject/verb)
2. N\V (object\verb)
3. A|N (adj|noun)
4. A|V (adj|verb)
5. N−N (noun−noun)
6. V−V (verb−verb)
7. A−A (add−adj)

opaque intrapropositional relations
8. V/V (infinitive_subject/verb)
9. V\V (infinitive_object\verb)
10. V|N (progressive_verb|noun, infinitive|noun)
11. A\V (prepositional_object\verb)

In addition, there are the extrapropositional relations (7.5.1). Of the 36 potential extrapropositional relations, English uses the following five:

7.6.5 EXTRAPROPOSITIONAL RELATIONS OF ENGLISH

12 $V/_xV$ sentential_subject$/_x$verb (opaque)
13 $V\backslash_xV$ sentential_object$\backslash_x$verb (opaque)
14 $V|_xN$ sentential_adnominal$|_x$noun, a.k.a. relative clause (opaque)
15 $V|_xV$ sentential_adverbial$|_x$verb (opaque)
16 $V-_xV$ extrapropositional$-_x$coordination (transparent)

Four elementary signatures are opaque (12–15) and one is transparent (16).
The relations 12, 13, 14, and 16 have intrapropositional counterparts in 7.6.4,
namely 8, 9, 10, and 6, respectively. We found altogether 8 transparent and 8
opaque elementary signatures in English.

The small number of altogether 16 intra- and extrapropositional elementary
signatures out of potentially 72 should make for easy language acquisition –
at least as far as the composition of content is concerned.[26] The binary nature
of elementary signatures is essential for the strictly time-linear, strictly surface
compositional processing of natural language.

As an example of a more complex construction consider the following content (see also 8.6.9):

7.6.6 CONTENT ANALYSIS CORRESPONDING TO A 20-WORD SENTENCE

(i) semantic relations graph (SRG)   *(iii) numbered arcs graph (NAG)*

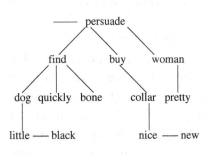

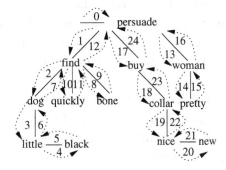

(ii) signature

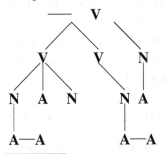

---

[26] Contents are essential for language acquisition insofar as they serve as the goal structure of the hear
mode and as the start structure of the speak mode.

(iv) surface realization

1	2	3	4	5–6	7	8	9–10	11–12
That	the	little	black	dog	found	the_bone	quickly	persuaded

13	14	15	16–17	18	19	20	21–22	23–24
the	pretty	woman	to_buy	a	nice	new	collar	.

All (1) subject/verb, (2) object\verb, (3) modifier|modified, and (4) conjunct−conjunct relations are characterized explicitly in the (i) signature and the (ii) semantic relations graph. The (iii) NAG provides the navigation order used for the (iv) surface realization. Even though the signature is semantic and has an initial node, it is not a semantic *hierarchy*.[27] Instead, the graph is a constellation of binary semantic relations of structure.[28]

The choice of which of two semantically related nodes must be the higher one in the graph is not determined by the parts of speech, as witnessed by the opaque constructions. Instead, it is determined by the semantic relation of functor-argument or coordination between the two nodes, which in turn constrains the possible choices of the parts of speech. This holds at the elementary, the phrasal, and the clausal level (3.5.6).

The SRG, the signature, and the NAG are alike in that they each have a unique entry point at the highest level. Furthermore, to get to an argument, the navigation must first traverse the functor; to get to a modifier, the navigation must first traverse the modified; and to get to a non-initial conjunct, the navigation must first traverse the initial conjunct.

For example, in 7.6.6 the extrapropositional navigation entering the proposition encounters the matrix verb first. From this entry point it proceeds to the subject, which happens to be a sentential argument. From the lower verb, the navigation continues through the first argument of the subject sentence, etc.

A navigation is driven by two complementary principles. The condition that all nodes should be traversed during a navigation (9.1.4, 4) is called the *downward principle*; it ensures that a new branch is entered if one is available. The condition that an intrapropositional navigation must end where it began (9.1.5, 2) is called the *upward principle*; it ensures that the navigation through a branch returns to the beginning. This enables the downward principle to apply again until all nodes have been traversed. Finally, the upward principle guides the navigation to the point of entry, i.e., the matrix verb.

Available branches at a given point in the navigation may be entered in different orders. This is shown by the comparison between the active 9.1.3 and the

---

[27] For example, there is no subclass or subtype relation between the V and the N in an N/V or any other elementary signature. For the treatment of subclass relations in DBS see Sect. 6.5.

[28] While every hierarchy may be represented as a graph, not every graph represents a hierarchy.

corresponding passive construction 9.1.7. Active traverses the NAG branches top down and from left to right. Passive traverses them top down and from right to left.

This alternative is reflected by the different numbering of arcs in the NAGs for active and passive. The numbering is purely descriptive, and forces the navigation to conform to a certain order for realizing a given content, represented by the SRG, in a certain language. There is another possibility, however, namely to let the navigation find the way through the graph on its own, based on the principle of *least effort*, and not on numbers which record the orders actually observed in a certain natural language example.

In DBS, the principle of least effort is applied (i) to traversing a branch completely or exiting early, (ii) to follow or not follow the time-linear order inherent in the content, and (iii) to not reversing or reversing the direction of the navigation. In short, a time-linear direction continuing in the same direction as long as possible constitutes a lesser effort than a multiple visit, an anti-time-linear direction, or a voluntary reversal of direction.

An example of exiting a branch early, requiring a multiple visit for a complete traversal, is unbounded dependency (9.4.2). An example of an anti-time-linear navigation is the traversal of an extrapositional coordination from right to left (9.2.4). A voluntary change of direction arises in the choice of passive over active.

Grammatical constructions which follow the principle of least effort are regarded as the linguistically unmarked[29] case in DBS, while constructions which run against least effort are regarded as marked. In this sense, realizing or interpreting an active, for example, requires less effort than realizing or interpreting the corresponding passive.

The reason that passive and the other marked traversals are used in communication anyway is due to other principles which override least effort. For example, theme-rheme (topic-comment), WH question formation, etc., induce word order requirements of their own which are often in marked contrast to the order corresponding to least effort.

For the linguist, the empirical challenge presented by a natural language example is the SRG, the signature, and the proplet representation; for this, the correct semantic details must be discovered, preferably for several different languages in parallel. The possible NAGs, in contrast, follow automatically.

For surface realization, the correct NAG must be selected. Also, the traversal steps which realize a surface must be indicated. They provide the input to LA-speak. Experience at the CLUE has shown that these tasks are easily acquired.

---

[29] See T. Givón (1990). Optimality Theory uses markedness for the ranking of constraints.

8. Simultaneous Amalgamation

A cognitive agent with language has the following sources of content: (i) current nonlanguage recognition, (ii) nonlanguage recognition stored in memory, (iii) inferences which derive new content from current or stored content, and (iv) language interpretation. These contents, regardless of their source, may be processed into blueprints which serve as input to the agent's action components (Chap. 5), including language production.

Thereby, language interpretation as a mapping from ordered surfaces to order-free content and language production as a mapping from order-free content to ordered surfaces are based on a strictly time-linear derivation order. However, no such order restriction needs to be assumed for nonlanguage recognition. Let us therefore define an LA-content grammar which applies the 16 elementary signatures defined in 7.6.4 and 7.6.5 as rules for the *simultaneous amalgamation* of content.

8.1 Intuitive Outline of LA-Content

Simultaneous amalgamation in nonlanguage recognition is based on strictly local and strictly binary relations between two elementary contents (nodes, proplets), as shown by the following example.

8.1.1 SIMULTANEOUS AMALGAMATION OF CONTENT CORRESPONDING TO ENGLISH The little black dog found the bone quickly.

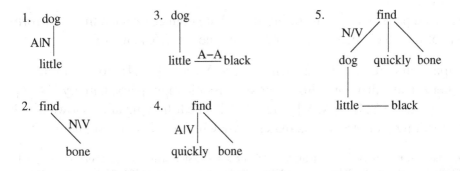

The hypothetical order assumed in 8.1.1 is determined by the order in which the raw nonlanguage data are being recognized and interpreted by the agent.

For example, the amalgamation might begin simultaneously with the A|N (*little*|*dog*) concatenation in 1 and the N\V (*bone**find*) concatenation in 2. Then these two separate content parts might each be extended simultaneously with the A−A (*little*−*black*) concatenation in 3 and the A|V (*quickly*|*find*) concatenation in 4, respectively. In 3, it is the first (lower) node of the A|N signature derived in 1 which is being extended. In 4 it is the second (upper) node of the N\V signature derived in 2 which is used for the extension. Finally, the top nodes of the two separate content parts 3 and 4 are concatenated in 5 with the N/V (*dog*/*find*) relation.

Any individual node, either isolated or already connected to other nodes, can be a local growth point in that it may be used as the first or second element of an elementary signature. For example, whether the nodes *dog* and *find* in concatenation 5 of 8.1.1 are already connected to other nodes (as shown in 3 and 4) or not does not matter as far as their local, binary concatenation is concerned. The main distinction in local concatenations is between obligatory connections, i.e., subject/verb and object\verb, and optional connections, i.e., conjunct−conjunct and modifier|modified (where the number and nature of the obligatory arguments are determined by the core value of the verb).

The amalgamation of nonlanguage content is based on the following steps:

8.1.2 STEPS OF AN ELEMENTARY AMALGAMATION

1. Raw data provided by the agent's visual, auditory, and other perception components are classified by concept types provided by the agent's memory, based on the principle of best match (cf. NLC'06, Sect. 4.4, and L&I'05) – as in a Rorschach test.
2. The instantiated concept tokens are embedded into N, V, or A proplet shells (6.6.1, 6.6.6).
3. Selected pairs of nodes resulting from 2 are connected such that they form one of the 16 elementary signatures[1] defined in 7.6.4 and 7.6.5.

The input pairs are unordered such that {N, V} and {V, N}, for example, are equivalent. Thus, there are altogether six possible input pairs, namely {N, V}, {N, A}, {V, A}, {N, N}, {V, V}, and {A, A}. The following table shows which of the resulting pair of nodes is matched by which elementary signature(s):

---

[1] Given the close relation between language and thought, it is assumed that the "parts of speech" and the elementary signatures of nonlanguage cognition are the same as those of the agent's native language.

8.1.3 UNAMBIGUOUS AND AMBIGUOUS INPUT TO SIGNATURES

Unambiguous match for {A, N}
3. A|N   (e.g., little|dog)

Unambiguous match for {N, N}
5. N−N   (e.g., man−woman)

Unambiguous match for {A, A}
7. A−A   (e.g., little−black)

Ambiguous match for {A, V}
4. A|V   (e.g., beautifully|sing)
11. A\V   (e.g., in_vase\put)

Ambiguous match for {N, V}
1. N/V   (e.g., John/gave)
2. N\V   (e.g., Mary\gave)
10. V|N   (e.g., burning|fire, to_help|desire)
14. V|$_x$N   (e.g., who_loves|Mary)

Ambiguous match for {V, V}
6. V−V   (e.g., walk−talk)
8. V/V   (e.g., to_err/is)
9. V\V   (e.g., to_read\try)
12. V/$_x$V   (e.g., that_bark/surprise)
13. V\$_x$V   (e.g., that_bark\hear)
15. V|$_x$V   (e.g., when_bark|smile)
16. V−$_x$V   (e.g., sleep−read)

The transparent adnominals and the nominal and adjectival coordinations are unique, while the other constellations show a growing number of ambiguities. These are the transparent adnominals and the nominal and adjectival coordinations, the transparent and the opaque adverbial, etc.[2]

The 16 elementary signatures allow us to connect any N, V, or A node to any other N, V, or A node. In the case of an unambiguous input pair, there is no need to choose between different elementary signatures for relating the two nodes. For an ambiguous input pair, in contrast, the choice between the possible elementary signatures must be based on additional information, mostly co-occurrence restrictions induced by the concepts involved (Sect. 8.5).

---

[2] This suggests that reuse is preferred over specialization. The more common, frequent, and obligatory the transparent semantic relation, the more the competing opaque relations riding piggy-back.

The LA-content grammar modeling simultaneous amalgamation must allow us to connect nodes in any derivation order. This is in contradistinction to the strictly time-linear derivation order of (i) LA-hear for the derivation of order-free content from language input, (ii) LA-think for the incremental activation of content, (iii) inferencing for the derivation of new order-free content, (iv) LA-speak for the derivation of surfaces from order-free content, and (v) LA-act for the derivation of nonlanguage action. The (vi) LA-content grammar maps free-order[3] nonlanguage recognition input into order-free content:[4]

8.1.4 INTERACTION OF FIVE COGNITIVE PROCEDURES

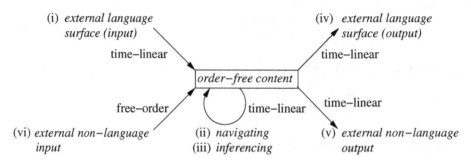

The free-order LA-content grammar (vi) is constrained insofar as it uses only N, A, and V as nodes, analyzed as three basic kinds of proplets, and only four kinds of connections, "/," "\," "|," and "−," resulting together in only 16 elementary semantic relations of structure (i.e., those listed in 7.6.4 and 7.6.5).

8.2 Formal Definition of LA-Content

Like all LA-grammars, LA-content defines a Start State ST_S, a set of rules, and a Final State ST_F. The 16 rules of LA-content realize the 16 elementary signatures of 7.6.4 and 7.6.5. A rule consists of two proplet patterns for matching input and two proplet patterns for deriving an elementary signature as output. An output differs from the input in that the output has a **prn** value and the core values have been cross-copied.[5]

---

[3] The term "free-order" is applied to the steps of a recognition procedure, while "order-free" is applied to items to be stored in the content-addressable memory of a Word Bank. Formally, a free-order recognition procedure based on an LA-content grammar can be easily narrowed into a procedure following a certain order, including partial orders. Thus, modeling an order, such as the saccades in the visual recognition of the mammal, may be accommodated.

[4] The assumption of an order-free content seems to agree with the SemR level of the Meaning-Text theory (MT) proposed by Zholkovskij and Mel'chuk (1965). An order-free level is also used in some Dependency Grammars, e.g., Hajicová (2000).

[5] The definition of LA-content in 8.2.1 benefitted from comments and suggestions by J. Handl.

8.2.1 DEFINITION OF LA-CONTENT

$ST_S =_{def} \{([RA: \alpha] \{rules\ 1\text{–}16\})\}$

Comment: The Start State of LA-content is unusual because it allows the initial proplet to have any *core attribute* (re-presented by the replacement variable for attributes RA), and any *core value* (re-presented by the binding variable α, cf. NLC'06, Sect. 4.1). Also, the rule package (hence RP) calls all of the 16 rules available. Next consider the first rule, N/V:

$$
1.\ \text{N/V:}\quad
\begin{bmatrix} \text{noun: } \alpha \\ \text{fnc:} \\ \text{prn:} \end{bmatrix}
\begin{bmatrix} \text{verb: } \beta \\ \text{arg:} \\ \text{prn:} \end{bmatrix}
\Rightarrow
\begin{bmatrix} \text{noun: } \alpha \\ \text{fnc: } \beta \\ \text{prn: K} \end{bmatrix}
\begin{bmatrix} \text{verb: } \beta \\ \text{arg: } \alpha \\ \text{prn: K} \end{bmatrix}
\quad \{\text{rules 2–6 and 8–16}\}
$$

Comment: The rule N/V corresponds to the intrapropositional subject-verb relation in English. The common prn value K indicates an intrapropositional relation. The relation is transparent because the attributes in the input patterns equal the attributes in the output patterns (this condition holds for the rules 1–7 and 16). The rule cross-copies the core values matched by α and β into the fnc slot of the noun and the arg slot of the verb. The rule package omits rule 1 (N/V), thus ensuring that an elementary proposition will have only one "subject" – though N−N coordinations of the subject are allowed. The RP omits rule 7 (A−A) because it cannot connect to N/V.

$$
2.\ \text{N\textbackslash V:}\quad
\begin{bmatrix} \text{noun: } \alpha \\ \text{fnc:} \\ \text{prn:} \end{bmatrix}
\begin{bmatrix} \text{verb: } \beta \\ \text{arg: X} \\ \text{prn:} \end{bmatrix}
\Rightarrow
\begin{bmatrix} \text{noun: } \alpha \\ \text{fnc: } \beta \\ \text{prn: K} \end{bmatrix}
\begin{bmatrix} \text{verb: } \beta \\ \text{arg: X } \alpha \\ \text{prn: K} \end{bmatrix}
\quad \{\text{rules 1–6, and 8–16}\}
$$

Comment: The rule N\V corresponds to the intrapropositional object-verb relation in English. Assuming X $\neq$ NIL, the only difference of this rule from rule 1 (N/V) is the *non-initial* position of α in the arg slot of the verb, indicating the role of an object. By its calling itself (rule 2 is included in the rule package), more than one object may be added; the number of objects required is determined "lexically" by the concept matching β. Allowing an application of N\V before N/V without permitting more than one subject requires somewhat more complicated input patterns, which are omitted for simplicity.

$$
3.\ \text{A|N:}\quad
\begin{bmatrix} \text{adj: } \alpha \\ \text{mdd:} \\ \text{prn:} \end{bmatrix}
\begin{bmatrix} \text{noun: } \beta \\ \text{mdr:} \\ \text{prn:} \end{bmatrix}
\Rightarrow
\begin{bmatrix} \text{adj: } \alpha \\ \text{mdd: } \beta \\ \text{prn: K} \end{bmatrix}
\begin{bmatrix} \text{noun: } \beta \\ \text{mdr: } \alpha \\ \text{prn: K} \end{bmatrix}
\quad \{\text{rules 1–2, 4–7, 10–11, and 14}\}
$$

Comment: The rule A|N corresponds to the intrapropositional relation between an adnominal modifier and a noun in English. The rule cross-copies core values matched by α and β into the mdd slot of the adj and the mdr slot of the noun. The RP omits A|N, but calls rule 7 (A−A); in this way multiple adnominal modifiers as in the fuzzy little black dog are treated as an intrapropositional coordination. The RP omits rules matching two V nodes because they cannot connect to A|N.

$$
4.\ \text{A|V:}\quad
\begin{bmatrix} \text{adj: } \alpha \\ \text{mdd:} \\ \text{prn:} \end{bmatrix}
\begin{bmatrix} \text{verb: } \beta \\ \text{mdr:} \\ \text{prn:} \end{bmatrix}
\Rightarrow
\begin{bmatrix} \text{adj: } \alpha \\ \text{mdd: } \beta \\ \text{prn: K} \end{bmatrix}
\begin{bmatrix} \text{verb: } \beta \\ \text{mdr: } \alpha \\ \text{prn: K} \end{bmatrix}
\quad \{\text{rules 1–4 and 6–16}\}
$$

Comment: The rule A|V corresponds to the intrapropositional relation between an adverbial modifier and a verb in English. The rule cross-copies the core values matched by α and β into the mdd slot of the adj and the mdr slot of the verb. The rule calls itself (the rule package contains A|V); in this way, a verb may have several modifiers, though not necessarily via a coordination. The RP omits rule 5 (N−N) because it cannot connect to A|V.

$$
5.\ \text{N−N:}\quad
\begin{bmatrix} \text{noun: } \alpha \\ \text{nc:} \\ \text{prn:} \end{bmatrix}
\begin{bmatrix} \text{noun: } \beta \\ \text{pc:} \\ \text{prn:} \end{bmatrix}
\Rightarrow
\begin{bmatrix} \text{noun: } \alpha \\ \text{nc: } \beta \\ \text{prn: K} \end{bmatrix}
\begin{bmatrix} \text{noun: } \beta \\ \text{pc: } \alpha \\ \text{prn: K} \end{bmatrix}
\quad \{\text{rules 1–3, 5, 10, 14}\}
$$

Comment: The rule N−N corresponds to the intrapropositional relation between two noun conjuncts in English. The rule cross-copies the core values matched by α and β into the nc slot of the first noun and the pc slot of the second noun. The RP calls rule 5 (N−N), and omits all rules which do not match an N, because they cannot connect to N−N.

6. V−V: $\begin{bmatrix} \text{verb: } \alpha \\ \text{nc:} \\ \text{prn:} \end{bmatrix}$ $\begin{bmatrix} \text{verb: } \beta \\ \text{pc:} \\ \text{prn:} \end{bmatrix}$ $\Rightarrow$ $\begin{bmatrix} \text{verb: } \alpha \\ \text{nc: } \beta \\ \text{prn: K} \end{bmatrix}$ $\begin{bmatrix} \text{verb: } \beta \\ \text{pc: } \alpha \\ \text{prn: K} \end{bmatrix}$ {rules 1–2, 4, 6, 8–16}

Comment: The rule V−V corresponds to the intrapropositional relation between two verb conjuncts in English. The rule cross-copies the core values matched by α and β into the nc slot of the first verb and the pc slot of the second verb. The RP calls V−V, and omits all rules which do not match a V, because they cannot connect to V−V, as in rule 8 (intrapropositional V/V), 9 (intrapropositional V\V), 12 (extrapropositional V$\Big/_x$V), 13 (extrapropositional V\$_x$V), and 16 (extrapropositional V−$_x$V).

7. A−A: $\begin{bmatrix} \text{adj: } \alpha \\ \text{nc:} \\ \text{prn:} \end{bmatrix}$ $\begin{bmatrix} \text{adj: } \beta \\ \text{pc:} \\ \text{prn:} \end{bmatrix}$ $\Rightarrow$ $\begin{bmatrix} \text{adj: } \alpha \\ \text{nc: } \beta \\ \text{prn: K} \end{bmatrix}$ $\begin{bmatrix} \text{adj: } \beta \\ \text{pc: } \alpha \\ \text{prn: K} \end{bmatrix}$ {3–4, 7, 11}

Comment: The rule A−A corresponds to the intrapropositional relation between two adnominal conjuncts in English. The rule cross-copies the core values matched by α and β into the nc slot of the first adjective and the pc slot of the second adjective. The RP calls rule 7 (A−A), and omits all rules which do not match an A, because they cannot connect to A−A.

8. V/V: $\begin{bmatrix} \text{verb: } \alpha \\ \text{arg:} \\ \text{prn:} \end{bmatrix}$ $\begin{bmatrix} \text{verb: } \beta \\ \text{arg:} \\ \text{prn:} \end{bmatrix}$ $\Rightarrow$ $\begin{bmatrix} \text{verb: } \alpha \\ \text{arg:} \\ \text{fnc: } \beta \\ \text{prn: K} \end{bmatrix}$ $\begin{bmatrix} \text{verb: } \beta \\ \text{arg: } \alpha \\ \text{prn: K} \end{bmatrix}$ {rules 2, 4, 6, 9–16}

Comment: The rule V/V corresponds to the intrapropositional relation between an infinitive serving as the subject and the matrix verb in English. The rule is opaque because a fnc attribute is added to the first pattern in the output. In this way, an infinitive serving as subject is connected to the matrix verb. The rule cross-copies the core values matched by α and β into the fnc slot of the first verb and the arg slot of the second verb. Rule 2 (N\V) in the rule package allows us to add a first object to the infinitive, as in to read a book, using the arg slot of the first verb. The RP omits all rules which do not match a V, because they cannot connect to V/V. RP also omits rule 1, rule 8 itself, and rule 12, ensuring that there can be only one subject.

9. V\V: $\begin{bmatrix} \text{verb: } \alpha \\ \text{arg:} \\ \text{prn:} \end{bmatrix}$ $\begin{bmatrix} \text{verb: } \beta \\ \text{arg: X} \\ \text{prn:} \end{bmatrix}$ $\Rightarrow$ $\begin{bmatrix} \text{verb: } \alpha \\ \text{arg:} \\ \text{fnc: } \beta \\ \text{prn: K} \end{bmatrix}$ $\begin{bmatrix} \text{verb: } \beta \\ \text{arg: X } \alpha \\ \text{prn: K} \end{bmatrix}$ {rules 1–2, 4, 6, 8–16}

Comment: The rule V\V corresponds to the intrapropositional relation between an infinitive serving as an object and the matrix verb in English. Assuming X $\neq$ NIL, the only difference of this rule from rule 8 (V/V) is the *non-initial* position of α in the arg slot of the second verb, indicating the role of an object.

10. V|N: $\begin{bmatrix} \text{verb: } \alpha \\ \text{arg:} \\ \text{prn:} \end{bmatrix}$ $\begin{bmatrix} \text{noun: } \beta \\ \text{mdr:} \\ \text{prn:} \end{bmatrix}$ $\Rightarrow$ $\begin{bmatrix} \text{verb: } \alpha \\ \text{mdd: } \beta \\ \text{prn: k} \end{bmatrix}$ $\begin{bmatrix} \text{noun: } \beta \\ \text{mdr: } \alpha \\ \text{prn: K} \end{bmatrix}$ {rules 1–6 and 8–16}

Comment: The rule V|N corresponds to the intrapropositional adnominal modifier relation between a progressive verb form and a noun in English, as in the burning fire. The rule is opaque because the arg attribute of the first input pattern is changed to mdd. The rule cross-copies the core values matched by α and β into the mdd attribute of the verb and the mdr attribute of the noun. The RP is similar to that of N/V and N\V; it omits rule 7 (A−A) because it can't connect to V|N.

11. A\V: $\begin{bmatrix} \text{adj: } \alpha \\ \text{mdd:} \\ \text{prn:} \end{bmatrix}$ $\begin{bmatrix} \text{verb: } \beta \\ \text{arg: X} \\ \text{prn:} \end{bmatrix}$ $\Rightarrow$ $\begin{bmatrix} \text{adj: } \alpha \\ \text{fnc: } \beta \\ \text{prn: K} \end{bmatrix}$ $\begin{bmatrix} \text{verb: } \beta \\ \text{arg: X } \alpha \\ \text{prn: K} \end{bmatrix}$ {rules 1–4 and 6–16}

Comment: The rule A\V corresponds to the relation between a prepositional object and its verb in English. The rule is opaque because the mdd attribute of the first input pattern is changed to fnc. The rule cross-copies the core values matched by α and β into the arg attribute of the verb and the fnc attribute of the adj. Assuming X $\neq$ NIL, a prepositional object can not assume the role of a subject. The RP is the same as in A|V; it omits rule 5 (N−N) because it can't connect to A\V.

12. $V/_xV$:
$\begin{bmatrix} \text{verb: } \alpha \\ \text{arg:} \\ \text{prn:} \end{bmatrix}$ $\begin{bmatrix} \text{verb: } \beta \\ \text{arg:} \\ \text{prn:} \end{bmatrix}$ $\Rightarrow$ $\begin{bmatrix} \text{verb: } \alpha \\ \text{arg:} \\ \text{fnc: } (\beta\ K+1) \\ \text{prn: } K \end{bmatrix}$ $\begin{bmatrix} \text{verb: } \beta \\ \text{arg: } (\alpha\ K) \\ \text{prn: } K+1 \end{bmatrix}$ {rules 1–2, 4, 6, 8–11, 13–16}

Comment: The rule $V/_xV$ corresponds to the relation between a subject sentence and its higher verb in English. The relation is extrapropositional, as indicated by the different **prn** values of the output. Therefore, the values resulting from cross-copying must include the **prn** value of the source proposition. The rule is opaque because the first input pattern is assigned an additional **fnc** attribute in the output. The rule cross-copies the core values matched by α and β into the **fnc** attribute of the first verb and the **arg** attribute of the second verb. The rule 1 (N/V) in the RP allows adding a subject to the first (lower) V, while the rules 2 (N\V), 11 (A\V), 13 (V\$_x$V), and 4 (A|V) allow adding a nominal, prepositional, or sentential object, and an adverbial modifier to either clause.

13. $V\backslash_xV$:
$\begin{bmatrix} \text{verb: } \alpha \\ \text{arg:} \\ \text{prn:} \end{bmatrix}$ $\begin{bmatrix} \text{verb: } \beta \\ \text{arg:} \\ \text{prn:} \end{bmatrix}$ $\Rightarrow$ $\begin{bmatrix} \text{verb: } \alpha \\ \text{arg:} \\ \text{fnc: } (\beta\ K+1) \\ \text{prn: } K \end{bmatrix}$ $\begin{bmatrix} \text{verb: } \beta \\ \text{arg: } X\ (\alpha\ K) \\ \text{prn: } K+1 \end{bmatrix}$ {rules 1–2, 4, 6, 8–16}

Comment: The extrapropositional rule $V\backslash_xV$ corresponds to the relation between an object sentence and its higher verb in English. Assuming $X \neq$ NIL, the only difference of this rule from rule 12 ($V/_xV$) is the *non-initial* position of α in the **arg** slot of the second verb, indicating the role of an object. The RP omits all rules which do not match a V, because they cannot connect to $V\backslash_xV$.

14. $V|_xN$:
$\begin{bmatrix} \text{verb: } \alpha \\ \text{arg:} \\ \text{prn:} \end{bmatrix}$ $\begin{bmatrix} \text{noun: } \beta \\ \text{mdr:} \\ \text{prn:} \end{bmatrix}$ $\Rightarrow$ $\begin{bmatrix} \text{verb: } \alpha \\ \text{arg:} \\ \text{mdd: } (\beta\ K) \\ \text{prn: } K+1 \end{bmatrix}$ $\begin{bmatrix} \text{noun: } \beta \\ \text{mdr: } (\alpha\ K+1) \\ \text{prn: } K \end{bmatrix}$ {rules 1–6 and 8–16}

Comment: The extrapropositional rule $V|_xN$ corresponds to the relation between a relative clause and its higher noun in English. The rule is opaque because the first output pattern is assigned an additional **mdd** attribute. The rule cross-copies the core values matched by α and β into the **mdd** attribute of the verb and the **mdr** attribute of the noun. The RP omits rule 7 (A–A), because it cannot connect to $V|_xV$, as in the rules 1 (N/V), 2 (N\V), and 10 (V|N)

15. $V|_xV$:
$\begin{bmatrix} \text{verb: } \alpha \\ \text{mdr:} \\ \text{prn:} \end{bmatrix}$ $\begin{bmatrix} \text{verb: } \beta \\ \text{mdr: } X \\ \text{prn:} \end{bmatrix}$ $\Rightarrow$ $\begin{bmatrix} \text{verb: } \alpha \\ \text{mdr:} \\ \text{mdd: } (\beta\ K+1) \\ \text{prn: } K \end{bmatrix}$ $\begin{bmatrix} \text{verb: } \beta \\ \text{mdr: } X\ (\alpha\ K) \\ \text{prn: } K+1 \end{bmatrix}$ {rules 1–2, 4, 6, 8–16}

Comment: The rule $V|_xV$ corresponds to the relation between an adverbial clause and its higher verb in English. The rule is opaque because the first input pattern is assigned an additional **mdd** attribute in the output. The rule cross-copies the core values matched by α and β into the **mdd** attribute of the first verb and the **mdr** attribute of the second verb. The RP omits all rules which do not match a V.

16. $V-_xV$:
$\begin{bmatrix} \text{verb: } \alpha \\ \text{nc:} \\ \text{prn:} \end{bmatrix}$ $\begin{bmatrix} \text{verb: } \beta \\ \text{pc:} \\ \text{prn:} \end{bmatrix}$ $\Rightarrow$ $\begin{bmatrix} \text{verb: } \alpha \\ \text{nc: } (\beta\ K+1) \\ \text{prn: } K \end{bmatrix}$ $\begin{bmatrix} \text{verb: } \beta \\ \text{pc: } (\alpha\ K) \\ \text{prn: } K+1 \end{bmatrix}$ {rules 1–2, 4, 6, 8–16}

Comment: The rule $V-_xV$ corresponds to an extrapropositional coordination in English. It is transparent and cross-copies the core values matched by α and β into the **nc** attribute of the first verb and the **pc** attribute of the second verb.

$ST_F =_{def} \{([\text{cat: } X]\ rp_{rules1--16}\}$
Comment: The Final State of the LA-content grammar includes all output states of rules 1–16.

Under normal circumstances (8.1.2), the agent's cognition will select (i) one concept interpretation for an elementary raw input, (ii) one proplet shell, and (iii) one elementary signature for relating two proplets resulting from (ii). However, as witnessed by the phenomenon of optical illusions, a given raw input may have more than one interpretation.

Applied to LA-content, this phenomenon makes it possible for a raw input to match (a) more than one concept, (b) more than one proplet shell, and/or (c) more than one elementary signature. Thereby, (a) and (b) correspond to a lexical ambiguity in natural language interpretation, and (c) corresponds to a syntactic-semantic ambiguity. What is the impact of these ambiguities on the complexity of LA-content defined in 8.2.1?

8.3 Linear Complexity of LA-Content

It is well-known that the only source of complexity in the class of constant LA-grammars (C-LAGs) is *recursive ambiguities* (TCS'92). The parsing of an input is recursively ambiguous if (i) it is recursive and (ii) the recursion repeats the same kind of ambiguity, resulting in a polynomial or exponential number of derivation steps (parallel readings).[6] Formal languages with recursive ambiguity are WW^R, accepting input like abcd...dcba (context-free), and WW, accepting input like abcd...abcd... (context-sensitive). In these two formal languages the recursive ambiguity arises in worst-case inputs like aaaaaa.[7]

C-LAGs for input without any recursive ambiguity form a sub-class called C1-LAGs and parse in linear time. The C1-LAGs recognize many context-sensitive languages, such as $a^k b^k c^k$, $a^k b^k c^k d^k$, and so on, $a^{2^i}, a^{n^2}, a^{i!}$, etc. Given the long history of attempting to fit natural languages into the Phrase Structure Hierarchy of complexity (cf. FoCL'99, Sects. 8.4 and 8.5), we would like to know how natural languages fit into the alternative complexity hierarchy of LA-grammar. This is equivalent to the empirical question of whether there exist natural language constructions which are not only ambiguous, but also recursively ambiguous (cf. FoCL'99, Sect. 12.5).

Recursive/iterative[8] structures are relatively numerous in natural language. Consider the following examples:

8.3.1 Recursive Structures in Natural Language

1. *Intrapropositional coordination*
   Examples: (i) The man, the woman, the child, ..., and the cat (noun co-ordination); cf. rule 5 (N−N) in 8.2.1. (ii) Peter bought, peeled, cooked,

---

[6] See FoCL'99, Sects. 11.3 and 11.4.

[7] According to the PSG hierarchy, the complexity of WW^R is context-free (n^3), while that of WW is context-sensitive (2^n or more; we know of no explicit Phrase Structure Grammar for this context sensitive language). In LA-grammar, in contrast, both are C2-languages (n^2 polynomial). The formal LA-grammars for WW^R and WW are defined in FoCL'99, 11.5.4 and 11.5.6, respectively.

[8] In computer science, every recursive function may be rewritten as an iteration, and vice versa. The choice between the two usually depends on the application and associated considerations of efficiency.

cut, spiced, served, ..., and ate the potatoes (verb coordination); cf.
rule 6 (V−V). (iii) The fuzzy clever little black hungry ... dog (adnominal coordination); cf. rule 7 (A−A).

2. *Extrapropositional coordination*
   Example: Julia slept. Susanne sang. John read.; cf. rule 16 (V−$_x$V)
   in 8.2.1, 3.2.5 for a proplet representation, and 7.4.4 for a DBS analysis.

3. *Iterated object sentences*
   Example: John said that Bill believes that Mary suspects that Suzy
   knows that Lucy loves Tom; cf. rule 13 (V\\$_x$V) in 8.2.1, and 9.4.1 for a
   DBS analysis. Related are the constructions of *unbounded* or *long distance
   dependency*, such as Who did John say that Bill believes that Mary
   suspects that Suzy knows that Lucy loves?, which are analyzed in
   9.4.2. Iterated object sentences may also serve as a subject sentence, as
   in That Bill believes that Mary suspects that Suzy knows that Lucy
   loves Tom surprised John.

4. *Iterated relative clauses*
   Example: The man who loves the woman who feeds the child who
   has a cat is sleeping; cf. rule 14 (V|$_x$N) in 8.2.1, and 9.3.3 for a DBS
   analysis.

5. *Gapping constructions*
   Examples: (i) Bob ate an apple, walked the dog, read the paper, had
   a beer, called Mary, ..., and took a nap. (subject gapping); cf. rule 1
   (N/V) in 8.2.1 and 9.5.5 for a DBS analysis. (ii) Bob ate an apple, Jim
   a pear, Bill a peach, Suzy some grapes, ..., and Tom a tomato. (verb
   gapping); cf. rules 1 (N/V) and 2 (N\\V), and 9.5.3 for a DBS analysis. (iii)
   Bob bought, Jim peeled, Bill sliced, Peter served, and Suzy ate the
   peach (object gapping); cf. rule 2 (N\\V), and 9.6.2 for a DBS analysis.
   (iv) Bob ate the red, the green, and the blue berries. (noun gapping);
   cf. rule 3 (A|N), and 9.6.4 for a DBS analysis.

6. *Iterated prepositional phrases*
   Example: Julia ate the apple on the table behind the tree in the gar-
   den ...; cf. rule 3 (A|N) in 8.2.1, and 7.2.4 for a partial DBS analysis.

Of these recursive constructions, only the last one could be construed as recursively ambiguous: each time a new prepositional phrase is added (recursion),
there is a systematic ambiguity between (i) an adnominal and (ii) an adverbial interpretation. By (needlessly) multiplying out the semantic readings in
the syntax, one may obtain exponential complexity. However, as shown in

FoCL'99, Sect. 12.5, this is simply a bad approach; there is the empirically adequate alternative of *semantic doubling*, which is of only linear complexity.

In short, LA-content may (i) contain recursive structures and (ii) have ambiguities, but still be a C1-LAG (if there are no recursive ambiguities). Whether or not LA-content as defined in 8.2.1 contains recursive ambiguities may be determined in two ways: (i) investigate the grammar by following the parallel possible continuations through the rule packages[9] or (ii) investigate the content phenomena which the grammar is designed to model.

If a recursive ambiguity is detected in the rule system, it may be empirically legitimate or not. If not, the recursive ambiguity must be eliminated from the grammar, for example, by modifying a rule package (debugging).

The first case, in contrast, would constitute an interesting linguistic discovery. However, as long as we can't find a recursive ambiguity as a phenomenon of natural language, and thus as a phenomenon of the corresponding content, LA-content must not exceed the linear complexity of the C1-languages – in concord with the CoNSyx hypothesis of FoCL'99, 12.5.7.

8.4 Infinitive Content Constructions

Let us turn now to a content construction which has not yet been treated in DBS, namely the infinitive. Its analysis will provide us with an opportunity to compare the assembling of content (i) by means of nonlanguage recognition (LA-content) and (ii) by means of natural language interpretation (LA-hear).

Good entry points for analyzing infinitive content constructions are corresponding surfaces of English.[10] At their center is a form of the verb, for example, see. Following the grammars for classical Latin, traditional grammars of English call it the infinitive, classify it as non-finite and unmarked for person, number, and tense,[11] and use it as the verb's base form ("citation form"). It should not be overlooked, however, that the surfaces of such infinitives as (to) see, (to) eat, and (to) give are the same as the unmarked finite non-third-person singular present tense forms in (I) see, (you) eat, and (they) give, respectively.

In many grammatical functions infinitives may be viewed as a stripped-down version of corresponding subclauses, e.g., a subject sentence or an object sen-

---

[9] Handl (2011) presents an algorithm which determines for any C-LAG whether it is recursively ambiguous or not. The algorithm runs in r^2, where r is the number of rules in the grammar.

[10] For an analysis of infinitive constructions in different languages, see Wurmbrandt (2001) and the review by Reis and Sternefeld (2004).

[11] We are referring here to an elementary verb form such as (to) see, leaving phrasal constructions such as (to) have seen or (to) have been seen aside.

tence (8.6.1). In DBS, this is formally expressed as follows: subclauses have a finite verb, which is reflected by their having their own **prn** value; infinitives, in contrast, borrow tense, mood, and agreement from their finite matrix verb, which is reflected by their sharing their **prn** value with the matrix clause.

Consider the following DBS hear mode derivation of the transitive infinitive construction **to read a book** serving as the object of the transitive verb **try**:

8.4.1 HEAR MODE DERIVATION OF **John tried to read a book**

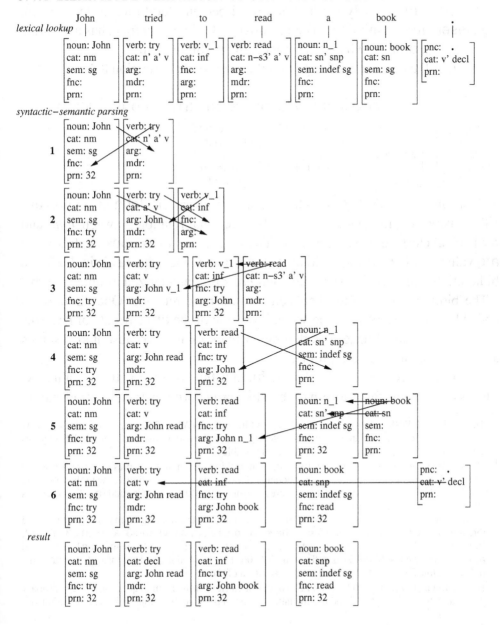

The lexical proplet representing **read** has the category (n-s3' a' v). It matches the RegEx pattern (n-s3' .* v), which represents the non-third-person singular present tense (i.e., unmarked) form of the English verb. The Kleene star * generalizes over varying oblique valency positions.[12] Because the surface of this form may be used systematically[13] as the infinitive, it would violate the principle of Surface Compositionality (SCG'84, FoCL'99, NLC'06)[14] to postulate separate lexical analyses for (i) the unmarked finite verb form of English and (ii) the English infinitive. The same holds for German, though it uses another finite verb form, namely the first and third person plural present tense form, e.g., **lesen**, for the infinitive. The infinitive in Latin, in contrast, has a separate surface, e.g., **amare**, which does not equal any finite form.

Consider the content resulting from the hear mode derivation 8.4.1:

8.4.2 CONTENT REPRESENTATION OF AN INFINITIVE CONSTRUCTION

$$
\begin{bmatrix} \text{noun: John} \\ \text{cat: nm} \\ \text{sem: sg} \\ \text{fnc: try} \\ \text{prn: 32} \end{bmatrix}
\begin{bmatrix} \text{verb: try} \\ \text{cat: decl} \\ \text{arg: John read} \\ \text{mdr:} \\ \text{prn: 32} \end{bmatrix}
\begin{bmatrix} \text{verb: read} \\ \text{cat: inf} \\ \text{fnc: try} \\ \text{arg: John book} \\ \text{prn: 32} \end{bmatrix}
\begin{bmatrix} \text{noun: book} \\ \text{cat: snp} \\ \text{sem: indef sg} \\ \text{fnc: read} \\ \text{prn: 32} \end{bmatrix}
$$

The proplet *John* has **try** as its **fnc** value, while *try* has **John** as its first **arg** value; using schema (1) in 7.3.3, the bidirectional relation between *John* and *try* is characterized as subject/verb. The proplet *read* has **book** as its second **arg** value, while *book* has **read** as its **fnc** value; using schema (2) in 7.3.3, the bidirectional relation between *read* and *book* is characterized as object\verb.

The bidirectional relation between *try* and *read* remains. One direction is coded by the value **read** in the second position of the **arg** attribute of *try*. This is opaque insofar as **read** is verbal rather than nominal because it also serves as the core value of a verb proplet.

The other direction is coded by the **fnc** attribute of *read* with the value **try**. This is opaque insofar as a verb proplet like *read* normally (i.e., in a transparent

---

[12] The equivalent DBS notation is (n-s3' X v). While early NEWCAT'86 implementations of LA-grammar used RegEx for matching ordered category values, the pattern matching for proplets requires a more differentiated method based on restricted variables.

[13] The single exception is the auxiliary **be** with its non-third-person singular present tense forms **am** and **are**. This exception is handled by introducing the variable INF (for infinitive) with the following restriction: INF ϵ {(n-s3' .* v), (be-inf)}. Thereby **be-inf** is the **cat** value of **be** and (n-s3' .* v) covers all other infinitives of English.

More specifically, the RegEx pattern (n-s3' .* v) matches the relevant forms of one-, two-, and three-place main verbs as well as those of **have** and **do**, i.e., (n-s3' v), (n-s3' a' v), (n-s3' d' a' v), (n-s3' hv' v), and (n-s3' do' v), respectively. In contrast, the forms **am** of category (ns1' be' v) and **are** of category (n-s13' be' v) as defined in NLC'06, 13.1.8, are not matched. Instead, the **cat** value of **be** is defined as **be-inf** and added to the restriction of the variable INF.

[14] Huddleston and Pullum (2002) postulate two lexical entries for the form in question, called "primary plain" and "secondary plain," and use the latter for the imperative, the subjunctive, and the infinitive.

construction) doesn't have a **fnc** attribute. In the LA-hear derivation 8.4.1, the **fnc** attribute is provided by the lexical *to* proplet, which receives the value **try** in line 2, before the *read* proplet is absorbed.

Because the object-verb relation between *to read* (infinitival object) and *try* (matrix verb) is opaque, it is not characterized by any of the transparent schemata in 7.3.3. Let us therefore define the following schema for the opaque object-verb relation V\V:

8.4.3 SCHEMA CHARACTERIZING AN ELEMENTARY V\V SIGNATURE

(9) verb\verb

$$\begin{bmatrix} \text{verb: } \beta \\ \text{fnc: } \alpha \\ \text{arg: } \gamma \text{ X} \end{bmatrix} \begin{bmatrix} \text{verb: } \alpha \\ \text{arg: } \gamma \beta \text{ Y} \end{bmatrix}$$

  to read   *try*   (examples of matching proplets for illustration only)

The number (9) corresponds to that of the corresponding elementary signature in 7.6.4.[15] The first verb pattern matches the infinitive serving as the object of the matrix verb. The variable β appears in the core attribute of the infinitive and in the object position of the **arg** slot of the matrix verb (second pattern), while the variable α appears in the core attribute of the matrix verb and in the **fnc** slot of the infinitive. Furthermore, the variable γ representing the subject of the matrix verb appears in the **arg** slot of the infinitive as subject, coding the implicit subject of the infinitive in the result of 8.4.1 (subject control).

Based on the proplet set 8.4.2 and the schemata 1 and 2 of 7.3.3 and 9 of 8.4.3, the procedure described in Sect. 7.3 will automatically derive (i) the SRG and (ii) the signature of the following DBS graph analysis:

8.4.4 DBS GRAPH ANALYSIS OF A CONTENT CORRESPONDING TO

John tried to read a book

(i) semantic relations graph (SRG)   *(iii) numbered arcs graph (NAG)*

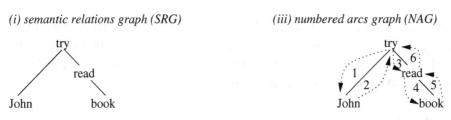

---

[15] The same holds for the numbering of the transparent schemata in 7.3.3, which also correspond to the numbers of the associated elementary signatures in 7.6.4.

(ii) signature

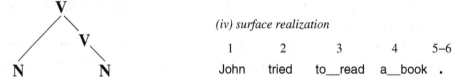

(iv) surface realization

1	2	3	4	5–6
John	tried	to__read	a__book	.

For language production, the indirect reconstruction of content based on the LA-hear derivation 8.4.1 must be complemented by a direct construction based on nonlanguage recognition and the LA-content grammar defined in 8.2.1. This may be illustrated by the following scenario:

> John and Mary are on a train in quiet conversation. At a station stop the train car is suddenly stormed by a large group of excited eleven-year-olds, making further talk impossible. Later Mary reports John's reaction as John tried to read a book.

The point is: How does Mary obtain the nonlanguage content required as input for this particular act of language production?

Let us replace Mary by an artificial agent, called AA. To obtain the content presented in 8.4.4, AA must apply its nonlanguage recognition to the raw data provided by the scene described. The purpose is to (i) establish the nodes *John, try, read,* and *book* as N, V, V, and N proplets, respectively, and (ii) to correctly establish the relations N/V, V\V, and N\V between them.

In 8.1.2 the search space of peripheral cognition (4.3.2) has already been reduced substantially with notions from central cognition. The burden on non-language recognition created by the ambiguities of the unordered {N, V}, {A, V}, and {V, V} input pairs (8.1.3) may be reduced further by utilizing the existing selectional constellations between relevant classes of concepts. This is the topic of the following section.

8.5 Selectional Constellations of Elementary Signatures

In infinitive content constructions, the relevant classes of concepts are the core values (i) in the matrix verb proplet and (ii) in the verb proplet of the infinitive. The class of matrix verbs like try, for example, may be established by enumerating the possible object constructions:

8.5.1 Try CLASS INFINITIVE CONSTRUCTIONS

1. nominal object: John tried a cookie.
2. one-place infinitive object: John tried to sleep.
3. two-place inf. object: John tried to read a book.

4. three-place inf. object: **John tried to give Mary a kiss.**
5. inf. with prepositional object: **Julia tried to put the flower in a vase.**
6. inf. with object sentence recursion: **Julia tried to say that Bill believes that Mary suspects that Susy knows that Lucy loves Tom. . . .**

The **try** class of concepts is defined as a two-place verb which can take (i) an infinitive or (ii) a noun as its object. The implicit subject of the infinitive equals the subject of the matrix concept (subject control). As indicated by the examples 2–6 in 8.5.1, the choice of the verb representing the infinitive is completely unrestricted in that it may be one-place, two-place, or three-place; taking a prepositional object, an object sentence, or an iteration of object sentences; and so on.

These properties may be formalized as the following schema:

8.5.2 DEFINITION OF **try** CLASS INFINITIVES

verb\verb   noun\verb

$$\begin{bmatrix} \text{verb: } \beta \\ \text{fnc: } \alpha \\ \text{arg: } \gamma\, X \end{bmatrix} \begin{bmatrix} \text{verb: } \alpha \\ \text{arg: } \gamma\, \beta \end{bmatrix} \qquad\qquad \begin{bmatrix} \text{noun: } \beta \\ \text{fnc: } \alpha \end{bmatrix} \begin{bmatrix} \text{verb: } \alpha \\ \text{arg: } \gamma\, \beta \end{bmatrix}$$

to read   *try*   *cookie*   *try*   (examples of matching proplets, for illustration only)

where $\alpha \in$ {begin, can afford, choose, decide, expect, forget, learn, like, manage, need, offer, plan, prepare, refuse, start, try, want}

Selectional constellations:

matrix verb α	*subject* γ	*infinitival object* β
begin 18 992	people 204, men 55, government 54, number 49, . . .	feel 528, be 492, take 371, . . .
can afford 1 841 . . .		. . .
. . .		

matrix verb α	*subject* γ	*nominal object* β
begin 6 642	government 32, people 26, commission 24, . . .	work 206, career 141, life 113, . . .
can afford 1 542 . . .		. . .
. . .		

As in the schema 8.4.3, the infinitive's wide range of possible objects is indicated by the variable X in the first schema (cf. [arg: γ X]).

Definition 8.5.2 refines 8.4.3 as follows. It (i) lists all possible matrix verb concepts of the **try** class as a restriction on the variable α. It (ii) provides an alternative schema for a noun object instead of an infinitive. And it (iii) lists the *selectional constellations* between each matrix verb α, its subjects γ and its infinitival or nominal objects β as n-tuples, for n=3. The numbers represent the actual BNC frequencies for each constellation, in decreasing order.[16]

---

[16] Thanks to T. Proisl for determining the BNC frequencies in 8.5.2 and 8.5.3. The restrictions on the variable α were gleaned from Sorensen (1997). In contradistinction to the *selectional restrictions* of generative grammar (e.g., Klima 1964), the *selectional constellations* of DBS record the distributional facts in a corpus. For the results to be compelling, an RMD corpus should be used.

The search space reduction provided by a definition like 8.5.2 works as follows: as soon as a first item in the raw data has been recognized by a certain concept, e.g., *book*, the total set of concepts and relations available to be matched onto the next raw data item is reduced to the set of those related to *book* in any way in the agent's Word Bank, beginning with the most frequent ones (cf. L&I'05). Like intersection (Sect. 5.4), this process converges quickly.

Thus, corpus linguistics may complement the work on an artificial agent (cf. Liu 2009a) even before robotic external interfaces are available. Using concept names as placeholders (6.6.8), the selectional constellations illustrated in 8.5.2 as well as the frequency values may be obtained from any given corpus.

The methods include word form recognition for (i) *lemmatization*, i.e., reduction to the base (citation) form, and (ii) for *categorization*, i.e., determining the morphosyntactic properties. They also include syntactic-semantic parsing for establishing the semantic relations of structure as listed in 7.6.4 and 7.6.5.

The relations between concepts found in a corpus may be embedded into a Word Bank already running with episodic content and world knowledge. This is possible because a Word Bank may be multidimensional (6.3.2) in that a given owner proplet may provide access to several token lines: their proplets must share the core value, but may otherwise serve different aspects. The Corpus Word Bank supports the agent's current speech production and interpretation.

The following example is restricted to the corpus aspect, and shows only the token lines for the owner record **decide** and its noun objects.

8.5.3 SORTING NOMINAL **decide** OBJECTS INTO A CORPUS WORD BANK

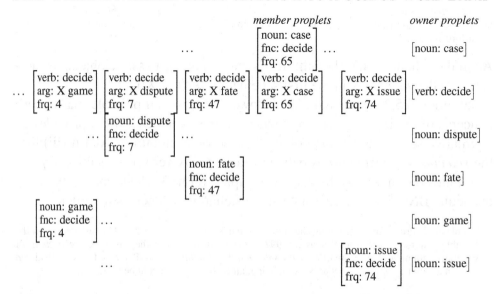

The owner proplets are ordered alphabetically, while the member proplets in a token line are ordered according to their **frq** (frequency) values. The example shows that **decide** occurred 65 times with the word **case** as object and that **case** occurred 65 time with the functor **decide** in the BNC corpus.

In contradistinction to the Episodic Word Bank 4.1.1, the Corpus Word Bank 8.5.3 is static. The content is changed only after more corpus work has revealed additional argument values for a given verb, additional functor values for a given noun, and so on, as well as different frequency numbers, such that the elements and their order in a token line have to be adjusted.

Querying a Corpus Word Bank is based on pattern proplets. The query pattern is matched with elements of a corresponding token line from right to left until a successful match is found (Sect. 4.2). For example, the question *How often does* **decide** *occur with the nominal object* **case** *in the BNC corpus?* may be formalized as the following query:

8.5.4 FORMAL QUERY AND ANSWER 1

query   *result*

$$
\begin{bmatrix} \text{verb: decide} \\ \text{arg: X case} \\ \text{frq: ?} \end{bmatrix}
\qquad
\begin{bmatrix} \text{verb: decide} \\ \text{arg: X case} \\ \text{frq: 65} \end{bmatrix}
$$

In the result, the "?" of the query is replaced by the **frq** value searched for.

Similarly, the question *How often does* **case** *occur as a nominal argument of* **decide** *in the BNC corpus?* may be formalized as the following query:

8.5.5 FORMAL QUERY AND ANSWER 2

query   *result*

$$
\begin{bmatrix} \text{noun: case} \\ \text{fnc: decide} \\ \text{frq: ?} \end{bmatrix}
\qquad
\begin{bmatrix} \text{noun: case} \\ \text{fnc: decide} \\ \text{frq: 65} \end{bmatrix}
$$

If **case** occurred also as the subject of **decide**, as in **This case decided the issue**, the query would return two answer proplets, one for **case** as subject, the other for it as object, each with its respective **frq** value.

Finally consider *Which intrapropositional relations in the BNC use* **decide***?*

8.5.6 FORMAL QUERY AND ANSWER 3

query   *result*

$$
\begin{bmatrix} \text{verb: decide} \\ \text{arg: ?} \\ \text{frq: ?} \end{bmatrix}
\quad
\begin{bmatrix} \text{noun: issue} \\ \text{fnc: decide} \\ \text{frq: 74} \end{bmatrix}
\begin{bmatrix} \text{noun: case} \\ \text{fnc: decide} \\ \text{frq: 65} \end{bmatrix}
\begin{bmatrix} \text{noun: fate} \\ \text{fnc: decide} \\ \text{frq: 57} \end{bmatrix}
\begin{bmatrix} \text{noun: dispute} \\ \text{fnc: decide} \\ \text{frq: 7} \end{bmatrix}
\begin{bmatrix} \text{noun: game} \\ \text{fnc: decide} \\ \text{frq: 4} \end{bmatrix}
$$

Here the query contains two question marks, and the return is the whole token line of **decide**.[17] In this query answer, the list starts with the most frequent.

8.6 Appear, Promise, and Persuade Class Infinitives

There are the following content structures corresponding to an infinitive construction in English:

8.6.1 CONTENT STRUCTURES CORRESPONDING TO INFINITIVES

1. Infinitive as subject: To err is human, to forgive divine.
2. Infinitive as object: John tried to read a book.
3. Infinitive as adnominal modifier: the desire to help
4. Bare infinitive:[18] Peter saw the accident happen.

(1) is largely restricted to copula constructions (e.g., 7.5.5 and 7.5.6). (2) has been illustrated in Sect. 8.4. (3) is a post-nominal variant of the V|N relation (7.6.4, 10). (4) has the following DBS graph analysis:

8.6.2 BARE INFINITIVE: Peter saw the accident happen.

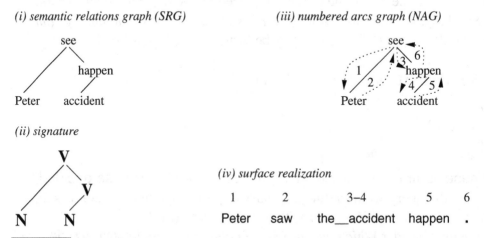

(i) semantic relations graph (SRG)

(iii) numbered arcs graph (NAG)

(ii) signature

(iv) surface realization

1	2	3–4	5	6
Peter	saw	the__accident	happen	.

---

[17] Selectional constellations for given natural languages are of special interest in machine translation (cf. Kay 1984, Dorr 1993, and Wu and Palmer 1994, among many others). Storing selectional constellations in a Corpus Word Bank may support this application by providing efficient and highly differentiated retrieval based on pattern proplets which utilize semantic relations.

[18] Bare infinitives are so called because they lack the **to**. Modal constructions, as in **Peter can see the horizon.**, do not reflect a separate kind of infinitive content construction and are treated like auxiliary constructions in DBS. Cf. FoCL'99, Sect. 17.3; NLC'06, Sect. 6.4.

Grammatically, **accident** is a direct object of the matrix verb **saw**, yet at the level of content it serves as the subject of the infinitive verb **happen** – which in turn functions as the real object of the matrix verb. As far as the surface order is concerned, the post-nominative rule for the position of the verb in English (cf. FoCL'99, Chap. 17), i.e. **accident happen(ed)**, seems to win over its other role as the object of the matrix clause, i.e., **saw happen**.

The resulting discrepancy between the grammatical appearance and the content construction resembles that of gapping (Sects. 9.5, 9.6): both constructions occur in many natural languages[19] with strong native speaker intuitions, but are rarely used.[20]

Each of the content constructions listed in 8.6.1 may consist of several classes. In the remainder of this section, we will consider different classes for infinitives as object. For example, in addition to including the **try** class discussed in Sects. 8.4 and 8.5, infinitives as object include the **appear** class. It resembles the **try** class defined in 8.5.2 in that it consists of concepts corresponding to transitive matrix verbs which take an infinitive as object. It differs, however, in that **appear** class matrix verbs do not allow nouns as objects.[21]

8.6.3 **Appear** CLASS INFINITIVE CONSTRUCTIONS

1. nominal object: *John appeared a cookie.
2. one-place infinitive object: John appeared to sleep.
   ...   (as in 8.5.1)

The illustrated properties may be formalized as the following schema:

8.6.4 DEFINITION OF **appear** CLASS INFINITIVES

verb\verb

$$
\begin{bmatrix} \text{verb: } \beta \\ \text{fnc: } \alpha \\ \text{arg: } \gamma \, \text{X} \end{bmatrix}
\begin{bmatrix} \text{verb: } \alpha \\ \text{arg: } \gamma \, \beta \end{bmatrix}
$$

to sleep   *appear*   (examples of matching proplets for illustration only)

where $\alpha \in$ {agree, appear, be able, seem, tend}

Selectional constellations: [omitted]

---

[19] The English bare infinitive corresponds to the Latin AcI, though with fewer selectional constellations.

[20] Thanks to T. Proisl for determining that 0.2% of the sentences in the BNC contain a bare infinitive.

[21] Exceptions are illustrated by examples like **John appeared five times** (frequency) and **John appeared on the scene** (locality). These nominal and prepositional objects are adverbial in character and illustrate once more the importance of recording and utilizing selectional constellations for each content word, based on a large RMD corpus (Reference Monitor corpus with Domain structure; cf. Sect. 12.2).

The restrictions on the variable α for **appear** and **try** class constructions (8.5.2) are disjunct.

A third content class of object infinitives is called the **promise** class. It differs from the **try** and **appear** classes in that the matrix verb has three valency positions, one for the subject, one for the indirect object, and one for the direct object. The latter may be either a noun or an infinitive.

8.6.5 Promise CLASS INFINITIVE CONSTRUCTIONS

1. nominal object: **John promised Mary a cookie.**
2. one-place infinitive object: **John promised Mary to sleep.**
   . . .   (as in 8.5.1)

Accordingly, the **promise** class is defined with the following two schemata:

8.6.6 DEFINITION OF promise CLASS INFINITIVES

$$\begin{bmatrix}\text{noun: } \delta \\ \text{fnc: } \alpha\end{bmatrix} \begin{bmatrix}\text{verb: } \beta \\ \text{fnc: } \alpha \\ \text{arg: } \gamma\, X\end{bmatrix} \begin{bmatrix}\text{verb: } \alpha \\ \text{arg: } \gamma\,\delta\,\beta\end{bmatrix} \qquad \begin{bmatrix}\text{noun: } \delta \\ \text{fnc: } \alpha\end{bmatrix} \begin{bmatrix}\text{noun: } \beta \\ \text{fnc: } \alpha\end{bmatrix} \begin{bmatrix}\text{verb: } \alpha \\ \text{arg: } \gamma\,\delta\,\beta\end{bmatrix}$$

Mary   to sleep   promise   Mary   cookie   promise (examples of matching proplets for illustration only)

where $\alpha \in$ {offer, promise, threaten[22]}

Selectional constellations:[omitted]

The schema on the left matches the elementary signatures N\V (indirect object) and V\V (infinitive), while the schema on the right matches transparent three-place verb constructions with the elementary signatures N\V (indirect object) and N\V (direct object).[23] The first noun pattern in either schema matches the indirect object, e.g., **Mary**. Its core value, represented by δ, appears in the second **arg** slot of the matrix verb α. The first value, γ, of this attribute represents the subject of the matrix verb and reappears in the subject slot of the infinitive verb β, thus coding subject control.

A fourth class of infinitive content construction is called the **persuade** class. It differs from the **promise** class in that the matrix verb may not take a noun instead of an infinitive. Also, the implicit subject of the infinitive is the nominal object of the matrix construction (object control). Consider the following examples:

---

[22] **Threaten** is an exception insofar as it doesn't seem to take a noun in place of the infinitive as its second object. This may be handled as part of the selectional constellations.

[23] The distinction between indirect and direct object is coded by the order of the **arg** values.

8.6.7 Persuade CLASS INFINITIVE CONSTRUCTIONS

1. nominal object: *John persuaded Mary a cookie.
2. one-place infinitive object: John persuaded Mary to sleep.
   ...   (as in 8.5.1)

Object control means that it is the object Mary who sleeps, in contrast to the examples in 8.5.1, in which it is the subject John.

While Mary is the indirect object in 8.6.5, it is the direct object in 8.6.7. This leaves the second arg position in the matrix verb to the infinitive.[24]

8.6.8 DEFINITION OF persuade CLASS INFINITIVES

$$\begin{bmatrix} \text{noun: } \delta \\ \text{fnc: } \alpha \end{bmatrix} \begin{bmatrix} \text{verb: } \beta \\ \text{fnc: } \alpha \\ \text{arg: } \delta\ \text{X} \end{bmatrix} \begin{bmatrix} \text{verb: } \alpha \\ \text{arg: } \gamma\ \beta\ \delta \end{bmatrix}$$

Mary   *to sleep*   *persuade*   (examples of matching contents, for illustration only)

where $\alpha \in$ {advise, allow, appoint, ask, beg, choose, convince, encourage, expect, forbid, force, invite, need, permit, persuade, select, teach, tell, urge, want, would like}.

Selectional constellations:[omitted]

As shown by the respective restrictions on the variable α, the number of verbs in the persuade class is substantially higher than that in the promise class (8.6.6).[25]

The schema matches the elementary signatures N\V (indirect object) and V\V (infinitive). The first noun pattern matches the direct object, e.g., Mary. Its core value, represented by δ, appears in the third arg slot of the matrix verb α and in the first arg slot of the infinitive β, thus coding object control.

In conclusion let us compare the DBS graph analysis of a persuade and a promise class infinitive content construction:

8.6.9 OBJECT CONTROL IN John persuaded Mary to read a book.

(i) semantic relations graph (SRG)   *(iii) numbered arcs graph (NAG)*

---

[24] Given that persuade class constructions do not allow replacing the infinitive by a nominal object, it does not necessarily follow that the infinitive in a persuade class construction must function as the indirect object of the matrix verb. In German, for example, there are three-place verbs like lehren (teach) which take two objects in the accusative (which would correspond to two direct objects in English).

[25] Comrie's (1986) claim that subject and object control are also determined by pragmatic factors may be accommodated by non-disjunct restriction sets on the variable α in various class definitions.

(ii) signature

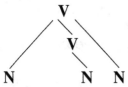

(iv) surface realization

1	2	3	4–5	6	7–8
John	persuaded	Mary	to__read	a__book	.

In the corresponding proplet representation, the implicit subject of the infinitive is specified explicitly as **Mary** in the **arg** slot of *read*:

8.6.10 CORRESPONDING PROPLET REPRESENTATION

$$
\begin{bmatrix} \text{noun: John} \\ \text{cat: nm} \\ \text{sem: sg} \\ \text{fnc: persuade} \\ \text{prn: 36} \end{bmatrix}
\begin{bmatrix} \text{verb: persuade} \\ \text{cat: decl} \\ \text{sem: past} \\ \text{arg: John read Mary} \\ \text{prn: 36} \end{bmatrix}
\begin{bmatrix} \text{noun: Mary} \\ \text{cat: nm} \\ \text{sem: sg} \\ \text{fnc: read} \\ \text{prn: 36} \end{bmatrix}
\begin{bmatrix} \text{verb: read} \\ \text{cat: inf} \\ \text{fnc: persuade} \\ \text{arg: Mary book} \\ \text{prn: 36} \end{bmatrix}
\begin{bmatrix} \text{noun: book} \\ \text{cat: snp} \\ \text{sem: indef sg} \\ \text{fnc: read} \\ \text{prn: 36} \end{bmatrix}
$$

The graphs of **promise** class constructions are similar to those of **persuade** class constructions except that the order of the infinitive and the nominal object is reversed. Consider the following example with a one-place infinitive:

8.6.11 SUBJECT CONTROL IN **John promised Mary to sleep.**

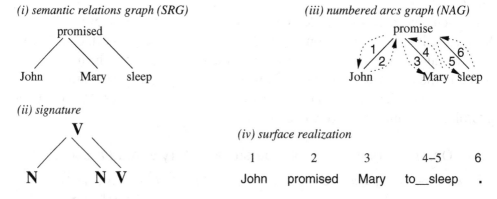

(i) semantic relations graph (SRG)

promised

John   Mary   sleep

(iii) numbered arcs graph (NAG)

promise

John   Mary   sleep

(ii) signature

V

N   N V

(iv) surface realization

1	2	3	4–5	6
John	promised	Mary	to__sleep	.

In the corresponding proplet representation, the implicit subject of the infinitive is specified explicitly as **John** in the **arg** slot of *sleep*:

8.6.12 PROPLET REPRESENTATION OF **John promised Mary to sleep.**

$$
\begin{bmatrix} \text{noun: John} \\ \text{cat: nm} \\ \text{sem: sg} \\ \text{fnc: promise} \\ \text{prn: 35} \end{bmatrix}
\begin{bmatrix} \text{verb: promise} \\ \text{cat: decl} \\ \text{sem: past} \\ \text{arg: John Mary sleep} \\ \text{prn: 35} \end{bmatrix}
\begin{bmatrix} \text{noun: Mary} \\ \text{cat: nm} \\ \text{sem: sg} \\ \text{fnc: promise} \\ \text{prn: 35} \end{bmatrix}
\begin{bmatrix} \text{verb: sleep} \\ \text{cat: inf} \\ \text{fnc: promise} \\ \text{arg: John} \\ \text{prn: 35} \end{bmatrix}
$$

In summary, the DBS class definitions for all kinds of constructions, including the infinitives, characterize complex contents by means of (i) schemata consisting of pattern proplets, (ii) explicit lists specifying the concepts which may serve as the matrix in the class schema (defined as restriction sets for the variable α), and (iii) an explicit listing of the selectional constellations between the concepts in a class schema, obtained from a corpus.

The DBS class schemata (e.g., 8.5.2, 8.6.4, 8.6.6, 8.6.8) constitute authentic linguistic generalizations, constrained empirically by the restriction set on the variable α and the n-tuples of selectional constellations. By storing the elements of such class definitions in an RMD Corpus Word Bank, they may be used to support lexical selection, for example, in machine translation. They may also be used to reduce the search space in artificial nonlanguage and language recognition.

The described method of determining the selectional constellations in a corpus depends crucially on the binary treatment of semantic relations in DBS. The method can not be used by systems computing possible substitutions such as Phrase Structure Grammar and Categorial Grammar.[26] This is because a substitution-based derivation of phrases and clauses is inherently compelled to use nonterminal nodes, resulting in non-binary (indirect) relations between the terminal nodes.

In a big corpus, the selectional constellations represent the content word distributions of many domains and many authors all mixed up. This is instructive for what might be regarded as the language as whole at a certain time. It is equally instructive, however, to analyze the content word distributions in a certain domain or in the work of a certain author.

Determining the selectional constellations for a big corpus, for a single domain, or for an author is based on the same routine in DBS. It consists of automatic word form recognition, syntactic-semantic parsing, frequency analysis, and storage in a Corpus Word Bank.

---

[26] As proven by C. Gaifman in June 1959, bidirectional Categorial Grammars and context-free Phrase Structure Grammars are weakly equivalent (Bar Hillel 1964, p. 103). See also Buszkowski (1988) and FoCL'99, Sect. 9.2.

9. Graph Theory

The binary nature of semantic relations in DBS makes them ideal for a graph-theoretical analysis: elementary signatures like N/V, N\V, A|N, etc. may be represented uniformly as two uninterpreted nodes (vertices) connected by an uninterpreted line (edge). In this way, the linguistically motivated distinctions between different parts of speech (nodes) and different kinds of semantic relations (lines) may be abstracted away from. What is focused on instead is the branching structure and the accessibility of nodes.

9.1 Content Analysis as Undirected and Directed Graphs

Central notions of graph theory are the *degree* of a node and the *degree sequence* of a graph. Consider, for example, the n=4 (four-node) graph known as K_4. K_4 is graph-theoretically *complete* because each node is connected to all the other nodes in the graph:

9.1.1 THE "COMPLETE" N=4 GRAPH K_4

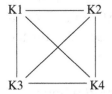

The degree sequence of this graph is 3333 because each of the four nodes is of degree 3 (i.e., connects three lines). The linguistically motivated signatures of 7.4.1, 7.4.4, and 7.4.5, in contrast, are incomplete. For example, disregarding any arc 0 line, the degree sequence of the signature in 7.4.1 is 2211, in 7.4.4 is 322111, and in 7.4.5 is 211.[1]

---

[1] The order of digits in a degree sequence begins with the highest degree, followed by the next highest degree, all the way down to the lowest degree nodes in the graph.

Let us compare K$_4$ with the following DBS graphs; they have the same number of nodes – and therefore the same number of digits in their respective degree sequences, but corresponding digits show different degrees:

9.1.2 SAME NUMBER OF NODES BUT DIFFERENT DEGREES

The little girl ate an apple.

(*i*) *semantic relations graph (SRG)*

eat

girl   apple

little

(*ii*) *signature*

V   2211

N   **N**

A

The man gave the child an apple

(*i*) *semantic relations graph (SRG)*

give

man   child apple

(*ii*) *signature*

V   3111

N   **N N**

The DBS graphs on the left have a two-place verb with its two arguments and the subject modified by an adnominal adjective, while the DBS graphs on the right have a three-place verb and its three arguments. Accordingly, the respective degree sequences are 2211 and 3111.

While the SRG and the signature in the DBS analysis of a content are *undirected* graphs, the associated numbered arcs graph (NAG) is a *directed* graph. Consider the following example in the active voice (see 9.1.7 for the corresponding passive):

9.1.3 DBS GRAPH ANALYSIS OF A CONTENT: The dog found a bone.

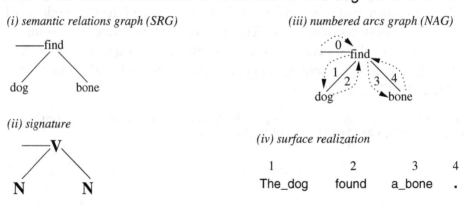

(*i*) *semantic relations graph (SRG)*

——find

dog   bone

(*ii*) *signature*

——**V**

N   **N**

(*iii*) *numbered arcs graph (NAG)*

0

find

1

2. 3 4

dog   bone

(*iv*) *surface realization*

1	2	3	4
The_dog	found	a_bone	.

A NAG may be based on an SRG, a signature, or a set of proplets (e.g., 7.4.2). The following constraints on NAGs apply only to intra- and extrapropositional functor-argument and intrapropositional coordination. Extrapropositional co-ordination requires a separate treatment, as shown in the following Sect. 9.2.

In order to be wellformed, a NAG must satisfy the following conditions:

9.1.4 GRAPH-THEORETICAL CONSTRAINTS ON WELLFORMED NAGS

1. The signature must be *simple*, i.e., there must be no loops or multiple lines.
2. The NAG must be *symmetric*, i.e., for every arc connecting some nodes A and B,[2] there must be an arc from B to A.
3. The traversal of arcs must be *continuous*, i.e., combining the traversal of arc x from A to B and of arc y from C to D is permitted only if B = C.
4. The numbering of arcs must be *exhaustive*, i.e., there must exist a naviga-tion which traverses each arc.

Because of conditions 1 (simple) and 2 (symmetric), the signature and the associated NAG are graph-theoretically *equivalent*.

However, because the numbering of the arcs induces a restriction on the traversal order of a NAG, the equivalence exists strictly speaking between the signature and the set containing all NAGs for the content in question.[3] As an example, consider the set of all possible NAGs corresponding to the n=3 (three-node) semantic relations graph of 9.1.3:

9.1.5 POSSIBLE NAGS FOR AN N=3 SEMANTIC RELATIONS GRAPH

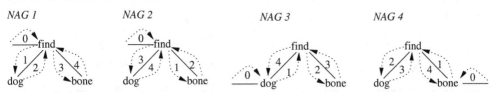

Graph-theoretically, the above graphs are all equally wellformed: excluding the start line, the arcs are symmetric, the numbering is continuous, and the traversal is exhaustive. Determining the possible NAGs for a given signature and the numbering of the arcs may be left to a simple algorithm.

For linguistic purposes not all four graphs in 9.1.5 are equally suitable, how-ever. In addition to graph-theoretical well-formedness (9.1.4), a NAG must also satisfy the following linguistic constraints:

---

[2] By requiring two nodes, a special exclusion of start lines is unnecessary.
[3] Alternatively, the equivalence exists between a signature and the one associated *un*numbered arcs graph. Or, if the traversal does not have to start at arc 1, the equivalence exists between a signature and each of its associated NAGs.

9.1.6 LINGUISTIC CONSTRAINTS ON WELLFORMED NAGS

1. An intrapropositional traversal must begin and end with the node which acts as the verb.[4]
2. The initial verbal node must be entered either by arc 0 or by a corresponding arc from a preceding proposition (7.4.4).
3. The only lines permitted in a NAG are "/" (subject-verb), "\" (object-verb), "|" (modifier-modified), and "−" (conjunct-conjunct).
4. The only nodes permitted in a NAG based on a signature are N (noun), V (verb), and A (adjective).

Condition 1 excludes the two NAGs on the right in 9.1.5. Because they do not begin and end with the verb, they cannot be coordinated with other propositions (cf. extrapropositional coordination 7.4.4 and 9.2.1).

The remaining two NAGs on the left in 9.1.5, in contrast, are linguistically wellformed. They are used for realizing the free word order of Russian (7.4.5) as well as the alternation between active and passive[5] in English:

9.1.7 DERIVING ENGLISH PASSIVE FROM CONTENT 9.1.3

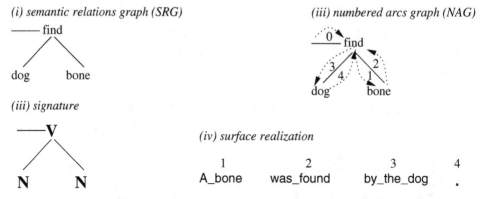

(i) semantic relations graph (SRG)

——— find

dog   bone

(iii) signature

———V

N   N

(iii) numbered arcs graph (NAG)

0 find

dog   bone

(iv) surface realization

1	2	3	4
A_bone	was_found	by_the_dog	.

The semantic relations and the signature of the passive are the same as in the active 9.1.3, while the NAG and the surface realization are different: the active uses the first NAG in 9.1.5, while the passive uses the second.

9.2 Extrapropositional Coordination

Extrapropositional coordination is bidirectional insofar as content may be traversed in the temporal and the anti-temporal direction (cf. NLC'06, Sect. 9.6). As an example, consider the following signature representing the content John leave house. John cross street.:

---

[4] Excepting contents without a verb (partial propositions), e.g., answers, certain headlines, etc.

9.2.1 EXTRAPROPOSITIONAL COORDINATION OF TWO PROPOSITIONS

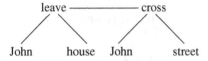

The corresponding proplet representation of this content connects the verb proplets *leave* and *cross* bidirectionally via their **nc** and **pc** slots:

9.2.2 BIDIRECTIONAL POINTERING IN THE PROPLET REPRESENTATION

$$
\begin{bmatrix} \text{noun: John} \\ \text{fnc: leave} \\ \text{prn: 1} \end{bmatrix}
\begin{bmatrix} \text{verb: leave} \\ \text{arg: John house} \\ \text{nc: (cross 2)} \\ \text{pc:} \\ \text{prn: 1} \end{bmatrix}
\begin{bmatrix} \text{noun: house} \\ \text{fnc: leave} \\ \text{prn: 1} \end{bmatrix}
\begin{bmatrix} \text{noun: John} \\ \text{fnc: cross} \\ \text{prn: 2} \end{bmatrix}
\begin{bmatrix} \text{verb: cross} \\ \text{arg: John street} \\ \text{nc:} \\ \text{pc: (leave 1)} \\ \text{prn: 2} \end{bmatrix}
\begin{bmatrix} \text{noun: street} \\ \text{fnc: cross} \\ \text{prn: 2} \end{bmatrix}
$$

The bidirectional connection at the proplet level (reflecting the corresponding signature) allows two kinds of traversal, forward and backward.

Forward navigation is based on the following NAG. It may be used for realizing the surface **John left the house. Then he crossed the street.**:

9.2.3 NAG FOR FORWARD NAVIGATION THROUGH 9.2.1

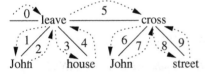

The corresponding backward navigation is based on the following NAG, which may be used for realizing the English surface **John crossed the street. Before that he left the house.**:

9.2.4 NAG FOR BACKWARD NAVIGATION THROUGH 9.2.1

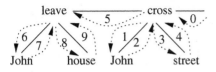

The point is that within a given traversal, e.g., 9.2.3 or 9.2.4, only one direction is used for the extrapropositional relation, i.e., the respective version of arc 5. This is different from the other semantic relations of structure, i.e., (i) intra- and (ii) extrapropositional functor-arguments as well as (iii) intrapropositional coordination, in which a complete traversal requires the use of both directions. Compare, for example, the extrapropositional coordination 9.2.3 with the following intrapropositional coordination:

---

[5] Cf. Twiggs (2005).

9.2.5 INTRAPROPOSITIONAL NOUN COORDINATION (SUBJECT)

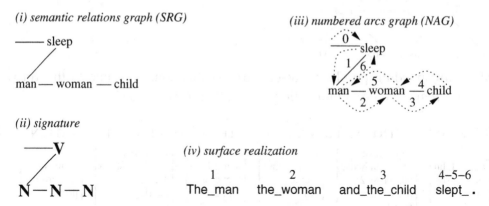

(i) semantic relations graph (SRG)

——— sleep

man — woman — child

(ii) signature

———V

N—N—N

(iii) numbered arcs graph (NAG)

0
sleep
1 6
5
man — woman — child
2   3
4

(iv) surface realization

1	2	3	4–5–6
The_man	the_woman	and_the_child	slept_ .

The NAG resembles that of an intrapropositional functor-argument, for example, 7.4.1, in that it is symmetric (each line has two arcs) and during realization each arc is traversed. More specifically, without arcs 4 and 5 in 9.2.5 it would be impossible to realize the full stop and to end the traversal at the initial verb node, as required by the linguistic constraint 1 in 9.1.6.

Next consider an extrapropositional functor-argument (7.5.1), here a subject sentence:

9.2.6 EXTRAPROPOSITIONAL FUNCTOR-ARGUMENT

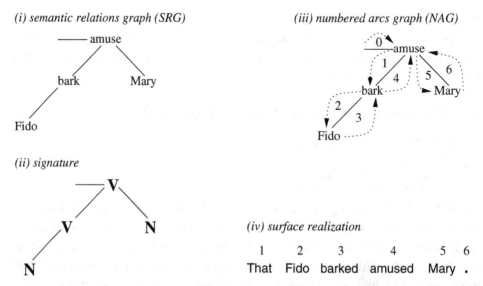

(i) semantic relations graph (SRG)

——— amuse

bark   Mary

Fido

(ii) signature

———V

V   N

N

(iii) numbered arcs graph (NAG)

0
amuse
1   4   5   6
bark   Mary
2
3
Fido

(iv) surface realization

1	2	3	4	5	6
That	Fido	barked	amused	Mary	.

Again, the NAG resembles that of an intrapropositional functor-argument, e.g., 7.4.1, in that it is symmetric and during realization each arc is traversed.

In short, extrapropositional coordination requires a treatment which is somewhat different from that of the other three grammatical structures. The basic choice is between the following graph-theoretical options:

9.2.7 GRAPH-THEORETICAL ALTERNATIVES FOR COORDINATION

1. unidirectional

2. bidirectional, symmetric

3. bidirectional, asymmetric

The unidirectional variant 1 has the advantage of not cluttering the representation with too many arcs. It is suitable for extrapropositional NAGs like 9.2.3 and 9.2.4, but not for intrapropositional coordination (9.2.5), and does not reflect adequately the concatenation structure of the associated proplets (9.2.2).

The bidirectional symmetric variant 2 reflects the possible backward navigation provided by the proplet representation. In 9.2.3 and 9.2.4 it may be added as an arc 10, defined as the inverse of the respective arc 5. Apart from arc 10 not being traversed in an extrapropositional navigation, variant 2 is equally suitable for intra- and extrapropositional coordination, and is like the other NAG structures by being symmetric as well.

The asymmetric bidirectional variant 3 is motivated by intrapropositional coordination. Consider the empty traversals of arcs 4 and 5 in 9.2.5: because there is no limit on the length of a coordination, the *empty traversals* on the return from an intrapropositional coordination require linear time. The alternative to variant 3 is to connect the backward arcs to the *initial* rather than the previous item.[6] This has the effect of reducing the cost of empty backward traversal from linear to constant. Variants 2 and 3 may be combined.

---

[6] In the proplet representation, this may be reflected by using an ic (for initial conjunct) rather than a pc (for previous conjunct) attribute. In a hear mode derivation, the value of the ic attribute is passed along incrementally, from conjunct n to conjunct $n+1$, making for simple rule patterns.

9.3 Sharing in Relative Clauses

Can our system as defined so far deal with difficult constructions of natural language, and by extension difficult constructions of content? As a case in point, let us consider center-embedded relative clauses of German.

Relative clauses function as sentential adnominal modifiers (Sect. 11.5), in contradistinction to adverbial clauses, which modify the verb (Sect. 11.6). As a brief introduction to relative clauses in DBS, compare the following SRGs of five relative clauses with their corresponding main clauses:

9.3.1 MAIN CLAUSES AND EQUIVALENT RELATIVE CLAUSES

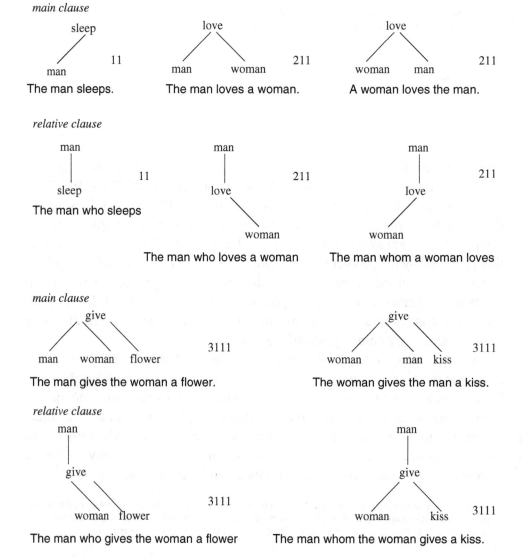

The degree sequence of a relative clause, written to the right of each graph, is the same as that of the corresponding main clause.

The semantically interpreted DBS graphs show that relative clauses use their modified (head nouns) as their subject, object, or prepositional argument. For example, in **The man who loves a woman** (subject gap) the graph shows no subject, just as in **The man whom the woman loves** (object gap) the graph shows no object. The "shared" noun, i.e., the modified one, is directly connected to the verb of the relative clause, which is no further than the distance between a noun and the verb in the corresponding main clause construction.

Turning to center-embedded relative clauses, we begin with an intuitive structural representation:

9.3.2 RELATIVE CLAUSE CENTER EMBEDDING

German: Der Mann   singt
                        der die Frau   liebt
                                        die das Kind füttert

Trans-
literation: the man   who the woman   who the child   feeds loves sings

Based on the meaning of the "/," "\," "|," and "−" lines, the DBS graph analysis of the center-embedded relative clause construction 9.3.2 turns out to be simple:

9.3.3 GRAPH ANALYSIS OF CENTER-EMBEDDED RELATIVE CLAUSES

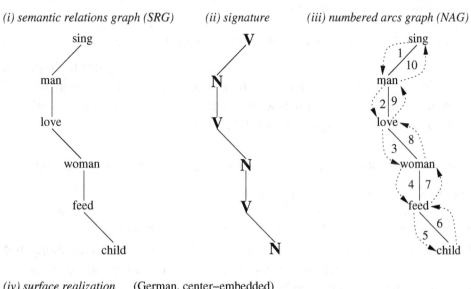

(i) semantic relations graph (SRG)   *(ii) signature*   *(iii) numbered arcs graph (NAG)*

(iv) surface realization   (German, center–embedded)

1	2	3	4	5	6	7–8	9–10
Der_Mann	der	die_Frau	die	das_Kind	fuettert	liebt	singt_ .

The DBS graph analysis has six nodes and the degree sequence 222211. The signature is given by our intuitions about functor-argument and coordination. These intuitions are language-independent in the sense that speakers of different languages can agree where two graph structures are alike and where they differ, for example, because of different lexicalization or different syntactic-semantic coding such as the *ergative*.[7]

The German surface realization walks down the left side of the NAG to realize the nouns, then back up on the right side to realize the verbs. *Multiple traversals* and *multiple realizations* (9.1.7) are evenly spread. Each arc is visited once and the traversal begins and ends with the main verb.

In comparison, consider the English counterpart:

9.3.4 ENGLISH REALIZATION OF CONTENT 9.3.3

The man who loves the woman who feeds the child sings.

Because English subclauses have the verb in post-nominative position (in contrast to the verb-final position in German subclauses), the first relative clause may be completed before the second begins. Therefore almost all nouns and verbs may be realized on the way down the left side of the NAG in 9.3.3:

9.3.5 ENGLISH SURFACE REALIZATION OF RELATIVE CLAUSES

surface realization   (English, unmarked)

1	2	3	4	5	6–7–8–9–10
The man	who_loves	the_woman	who_feeds	the_child	sings_ .

Going back up on the right side of the NAG, there is nothing left to do except at the very end (arc 10), when the main verb and the full stop are realized.

Another possibility of English word order is the *extraposition* of a relative clause, as in the following variant of 9.3.5:

9.3.6 SURFACE REALIZATION WITH EXTRAPOSED RELATIVE CLAUSE

surface realization   (English, extraposed, marked)

1	10	1–2	3	4	5	6–7–8–9–10
The_man	sings	who_loves	the_woman	who_feeds	the_child	.

The realization of this surface requires a *multiple visit*. After realizing **the man** in arc 1, the navigation returns to the verb and realizes **sings** in arc 10. From there, there is no choice but to traverse arc 1 again (multiple visit) and continue with arc 2 to the relative clause. Then the navigation travels down the

---

[7] Thanks to the many native speakers at the CLUE who participated in these enlightening discussions.

left side of the NAG in 9.3.3, realizing the verbs and the nouns. On the way back up, there is nothing left to do except to realize the full stop.

The surface realization 9.3.6 shows that a consecutive numbering is a sufficient, but not a necessary, condition 3 for satisfying continuity (9.1.4). For example, the combined traversal of arcs 1 and 10 is continuous even though the arc numbers are not consecutive.

From the software side, the mechanism of multiple visits is easily programmed. From the cognitive and linguistic sides, however, multiple visits must be constrained because otherwise there is no limit on complexity. The constraint applies to connected graphs and comes in two variants.

9.3.7 CONSTRAINT ON MULTIPLE VISITS, VARIANT I

In content navigation without language realization, a multiple visit is
1. *permitted* if there are still untraversed arcs in the graph,
2. *prohibited* if all arcs in the graph have been traversed.

These conditions are based on the possibility of keeping track of how often a node has been traversed. Constraint I may be relaxed by setting limits on how often a node may be traversed (traversal counter; cf. FoCL'99, p. 464).

9.3.8 CONSTRAINT ON MULTIPLE VISITS, VARIANT II

In content navigation with language realization, a multiple visit is
1. *permitted* if a required function word has not yet been realized,
2. *prohibited* if there is no value remaining in the current proplet set which has not already been used exhaustively for realization.

Technically, variant II is based on a language-dependent LA-speak grammar for mapping content into surfaces.

9.4 Unbounded Dependencies

In the previous section, the same NAG 9.3.3 was used for realizing the surface (i) of a non-extraposed English relative clause (unmarked order; 9.3.5) and (ii) of the corresponding extraposed relative clause (marked order; 9.3.6). This method, based on multiple visits for realizing the marked order, may also be applied to a construction of English known as *unbounded dependency* or *long distance dependency*.

The unmarked order of this construction is an iteration of object sentences, as in the following example:

9.4.1 DBS GRAPH ANALYSIS FOR
John said that Bill believes that Mary loves Tom.

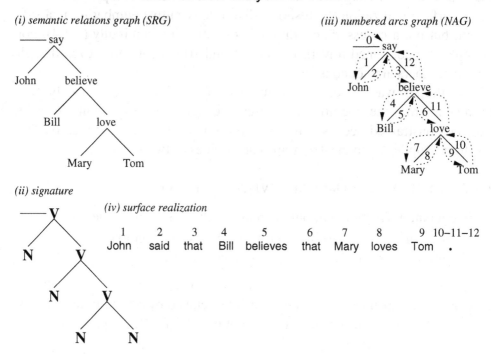

(i) semantic relations graph (SRG)

(iii) numbered arcs graph (NAG)

(ii) signature

(iv) surface realization

	1	2	3	4	5	6	7	8	9	10–11–12
	John	said	that	Bill	believes	that	Mary	loves	Tom	.

The iteration may be continued indefinitely, for example, by replacing the last object sentence with that Mary suspects that Suzy loves Tom or that Mary suspects that Suzy knows that Lucy loves Tom.

Next consider the corresponding unbounded dependency construction:

9.4.2 NAG AND SURFACE REALIZATION OF
Whom did John say that Bill believes that Mary loves?

(iii) numbered arcs graph (NAG)

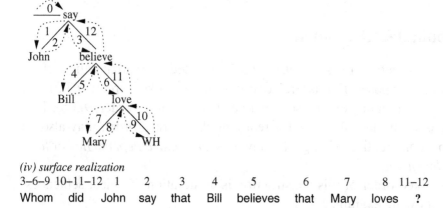

(iv) surface realization

3–6–9 10–11–12	1	2	3	4	5	6	7	8 11–12
Whom	did	John	say	that	Bill	believes	that	Mary	loves	?

The semantic relations and signature graphs of the iterated object sentence construction 9.4.1 and the corresponding unbounded dependency 9.4.2 are the same; both have the degree sequence 2221111 (disregarding the start line). The only difference between the respective NAGs is in the object of the final object sentence, i.e., **Tom** or **WH**. The dependency in 9.4.2 is between the surface-initial WH and the object of the final object sentence; it is unbounded because there is no limit on the number of intervening iterated object sentences.

The order of proplets in the following content representation of 9.4.2 follows the NAG. Given that proplets are inherently order-free, they could just as well be shown in the alphabetical order of the core values or in the surface order:

9.4.3 PROPLET REPRESENTATION OF 9.4.2

$$
\begin{bmatrix} \text{verb: say} \\ \text{arg: John (believe 7)} \\ \text{prn: 6} \end{bmatrix}
\begin{bmatrix} \text{noun: John} \\ \text{fnc: say} \\ \text{prn: 6} \end{bmatrix}
\begin{bmatrix} \text{verb: believe} \\ \text{arg: Bill (love 8)} \\ \text{fnc: (say 6)} \\ \text{prn: 7} \end{bmatrix}
\begin{bmatrix} \text{noun: Bill} \\ \text{fnc: believe} \\ \text{prn: 7} \end{bmatrix}
\begin{bmatrix} \text{verb: love} \\ \text{arg: Mary WH} \\ \text{fnc: (believe 7)} \\ \text{prn: 8} \end{bmatrix}
$$

$$
\begin{bmatrix} \text{noun: Mary} \\ \text{fnc: love} \\ \text{prn: 8} \end{bmatrix}
\begin{bmatrix} \text{noun: WH} \\ \text{fnc: love} \\ \text{prn: 8} \end{bmatrix}
$$

The verb proplet *say* has the arguments **John** and **(believe 7)**, the verb proplet *believe* has the arguments **Bill** and **(love 8)**, and the verb proplet *love* has the arguments **Mary** and **WH**.

The verb proplets *believe* and *love* of the object sentences are opaque (Sect. 7.5) because they have an additional **fnc** attribute to serve as an argument of their higher clauses. The values of these **fnc** attributes are addresses, specified as the core and the **prn** value of the preceding higher verb, just as the **arg** attributes of the higher verbs specify the following subclause as an address (bidirectional pointering).

Finally consider the surface realization of the unbounded dependency construction based on the NAG in 9.4.2. The WH word is accessed and realized by traversing arcs 3-6-9. This navigation is continuous, in line with 9.1.4, 3, despite the nonconsecutive numbering. Then the auxiliary **did** is realized by returning to *say* via arcs 10-11-12, relying on the usual *do-support* of English WH constructions. From there, the realization from arc 1 to arc 8 resembles that of the iterated object sentence construction 9.4.1. Omitting arcs 9 and 10, the navigation returns to *say* via 11-12 to realize the fullstop.[8]

---

[8] Superficially, the signature in 9.4.1 may seem to resemble a Nativist Phrase Structure tree. However, as explained in Sect. 7.1, the lines in a Phrase Structure tree indicate dominance and precedence, whereas the lines in a DBS graph indicate the subject/verb, the object\verb, the modifier|modified, and the conjunct−conjunct relations.

Furthermore, Nativism handles unbounded dependencies by "fronting" the WH word, thus chang-

9.5 Subject Gapping and Verb Gapping

A veritable pinnacle of grammatical complexity is *gapping* (cf. Kapfer 2010). In contrast to sentential arguments and modifiers, which are extrapropositional (7.5.1, Sects. 11.3–11.6), gapping is an intrapropositional[9] construction: one or more gapping parts share a subject, a verb, or an object with a complete sentence; in addition, there is noun gapping, in which several adnominals share a noun.

In common practice, gapping is rarely used.[10] Yet native speakers from a wide range of languages[11] confidently confirm that the various forms of gapping exist in their native tongue. This allows only one conclusion: gapping must be a very basic if not primitive construction which closely mirrors the underlying thought structure.

From the time-linear viewpoint of DBS, the most basic structural distinction between different kinds of gapping in English is whether the filler precedes or follows the gap(s). Subject and verb gapping have in common that the filler comes first. Consider the following examples:

9.5.1 SUBJECT GAPPING

Bob ate an apple, # walked the dog, and # read the paper.

The subject of the complete sentence, **Bob**, is shared by gapped parts which follow, with # marking the gaps.[12]

9.5.2 VERB GAPPING

Bob ate an apple, Jim # a pear, and Bill # a peach.

The verb of the complete sentence, **ate**, is shared by gapped parts which follow, with # marking the gaps.[13]

---

ing the tree. DBS, in contrast, leaves the graph unchanged and handles the different word orders of the iterated object sentence and the corresponding unbounded dependency by means of alternative navigations through the same NAG.

[9] The main parts of an extrapropositional construction have different **prn** values, while in an intrapropositional construction all proplets share the same **prn** value.

[10] März (2005) estimates that only 0.25% of the sentences in the LIMAS corpus contain gapping constructions.

[11] They include Albanian, Bulgarian, Chinese, Czech, English, French, German, Georgian, Italian, Japanese, Korean, Polish, Romanian, Russian, Spanish, Swedish, and Tagalog.

[12] Less than 0.3% of the sentences in the BNC show this construction of subject gapping. Thanks to T. Proisl, who provided the BNC frequencies for this and the following gapping constructions.

[13] Less than 0.01% of the sentences in the BNC show this construction of verb gapping.

Motivated by the presence of **and**, the gapping analysis in NLC'06 assumed an intrapropositional coordination structure. This worked well for subject and object gapping, in which a coordination may be defined between the different verbs contained in the gapping parts. But what if the gapping parts do not contain any verb because it is the verb that is gapped?

When we finally subjected verb gapping to a DBS graph analysis, the following solution was clear to see:

9.5.3 DBS GRAPH ANALYSIS OF VERB GAPPING IN 9.5.2

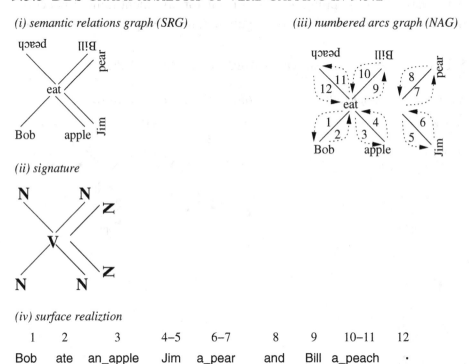

(i) semantic relations graph (SRG)

(iii) numbered arcs graph (NAG)

(ii) signature

(iv) surface realiztion

1	2	3	4–5	6–7	8	9	10–11	12
Bob	ate	an_apple	Jim	a_pear	and	Bill	a_peach	.

For graphical reasons, the lines representing the subject verb and the object verb relations in the gapping parts **Jim a pear** and **Bill a peach** are oriented properly only when rotated (see the orientation of the writing). The n=7 signature has the degree sequence 6111111. The surface realization uses each arc in the NAG once (no multiple visits), with empty traversals in 4, 6, and 10, and double realizations in 3, 7, and 11.

The graph analysis 9.5.3 requires a proplet representation different from the coordination-oriented solution of NLC'06, Sect. 8.5. Here the semantic relations between the gapping constructions and the filler are run via the **arg** attribute of the lone verb and the **fnc** attributes of the nouns in the gapping constructions – and not via their **nc** and **pc** attributes.

9.5.4 VERB GAPPING CONTENT AS A SET OF PROPLETS

$$
\begin{bmatrix} \text{noun: Bob} \\ \text{cat: nm} \\ \text{sem: sg} \\ \text{fnc: eat} \\ \text{prn: 31} \end{bmatrix}
\begin{bmatrix} \text{verb: eat} \\ \text{cat: decl} \\ \text{sem: past} \\ \text{arg: Bob apple} \\ \quad \text{Jim pear} \\ \quad \text{Bill peach} \\ \text{prn: 31} \end{bmatrix}
\begin{bmatrix} \text{noun: apple} \\ \text{cat: snp} \\ \text{sem: indef sg} \\ \text{fnc: eat} \\ \text{prn: 31} \end{bmatrix}
\begin{bmatrix} \text{noun: Jim} \\ \text{cat: nm} \\ \text{sem: sg} \\ \text{fnc: eat} \\ \text{prn: 31} \end{bmatrix}
\begin{bmatrix} \text{noun: pear} \\ \text{cat: snp} \\ \text{sem: indef sg} \\ \text{fnc: eat} \\ \text{prn: 31} \end{bmatrix}
\begin{bmatrix} \text{noun: Bill} \\ \text{cat: nm} \\ \text{sem: sg} \\ \text{fnc: eat} \\ \text{prn: 31} \end{bmatrix}
\begin{bmatrix} \text{noun: peach} \\ \text{cat: snp} \\ \text{sem: indef sg} \\ \text{fnc: eat} \\ \text{prn: 31} \end{bmatrix}
$$

The sharing of the verb is expressed by listing the three subject-object pairs in the **arg** attribute of the verb proplet and the verb in the **fnc** slot of the nouns.

The analysis of verb gapping in terms of functor-argument rather than coordination may be applied to all forms of gapping, for example, subject gapping:

9.5.5 DBS GRAPH ANALYSIS OF SUBJECT GAPPING IN 9.5.1

(i) semantic relations graph (SRG)   *(iii) numbered arcs graph (NAG)*

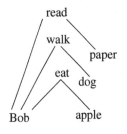   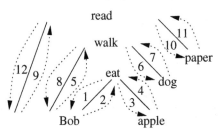

(iii) signature

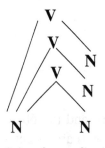

(iv) surface realization

1	2	3	4–1–5	6	7–8–9	10	11–12–2
Bob	ate	an_apple	walked	the_dog	and_read	the_paper	.

The degree sequence is 3222111. In contradistinction to verb gapping 9.5.3 and object gapping 9.6.1, subject gapping 9.5.5 shows multiple visits, namely 4-1-5 and 11-12-2.

The graphical analysis 9.5.5 requires a proplet representation different from the coordination-oriented solution of NLC'06, Sect. 8.4, as well:

9.5.6 Proplet Representation of Subject Gapping

$$
\begin{bmatrix} \text{noun: Bob} \\ \text{cat: nm} \\ \text{sem: sg} \\ \text{fnc: eat} \\ \qquad \text{walk} \\ \qquad \text{read} \\ \text{prn: 32} \end{bmatrix}
\begin{bmatrix} \text{verb: eat} \\ \text{cat: decl} \\ \text{sem: past} \\ \text{arg: Bob apple} \\ \text{prn: 32} \end{bmatrix}
\begin{bmatrix} \text{noun: apple} \\ \text{cat: snp} \\ \text{sem: indef sg} \\ \text{fnc: eat} \\ \text{prn: 32} \end{bmatrix}
\begin{bmatrix} \text{verb: walk} \\ \text{cat: decl} \\ \text{sem: past} \\ \text{arg: Bob dog} \\ \text{prn: 32} \end{bmatrix}
\begin{bmatrix} \text{noun: dog} \\ \text{cat: snp} \\ \text{sem: def sg} \\ \text{fnc: walk} \\ \text{prn: 32} \end{bmatrix}
\begin{bmatrix} \text{verb: read} \\ \text{cat: decl} \\ \text{sem: past} \\ \text{arg: Bob paper} \\ \text{prn: 32} \end{bmatrix}
\begin{bmatrix} \text{noun: paper} \\ \text{cat: snp} \\ \text{sem: def sg} \\ \text{fnc: read} \\ \text{prn: 32} \end{bmatrix}
$$

Here the shared item is the subject *Bob*, which lists the three verbs in its fnc slot, just as the three verbs list Bob in the first position of their arg slot.

9.6 Object Gapping and Noun Gapping

While English subject gapping and verb gapping have in common that the filler precedes the gap(s), object gapping and noun gapping have in common that the filler follows the gap(s). Consider the following example of object gapping:

9.6.1 Object Gapping:

   Bob bought #, Jim peeled #, and Bill ate the apple.

The object of the complete sentence, **apple**, is shared by the gapped parts preceding, with # marking the gaps.[14] The DBS graph analysis is as follows:

9.6.2 DBS Graph Analysis of Object Gapping in 9.6.1

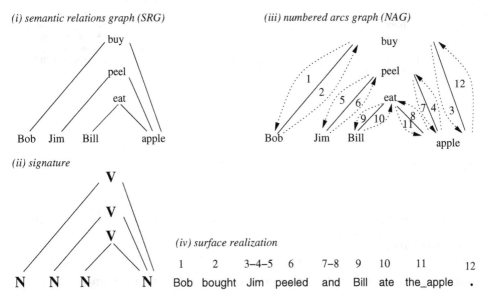

(i) semantic relations graph (SRG)

(iii) numbered arcs graph (NAG)

(ii) signature

(iv) surface realization

	1	2	3–4–5	6	7–8	9	10	11	12
	Bob	bought	Jim peeled	and	Bill	ate	the_apple	.	

---

[14] Less than 0.01% of the sentences in the BNC show this construction of object gapping.

The degree sequence is the same as in subject gapping (9.5.5), namely 3222111. Again, the revised graphical analysis of object gapping requires a proplet representation different from the coordination-oriented approach of NLC'06, Sect. 8.6:

9.6.3 PROPLET REPRESENTATION OF OBJECT GAPPING

$$
\begin{bmatrix} \text{noun: Bob} \\ \text{cat: nm} \\ \text{sem: sg} \\ \text{fnc: buy} \\ \text{prn: 33} \end{bmatrix}
\begin{bmatrix} \text{verb: buy} \\ \text{cat: decl} \\ \text{sem: past} \\ \text{arg: Bob apple} \\ \text{prn: 33} \end{bmatrix}
\begin{bmatrix} \text{noun: Jim} \\ \text{cat: nm} \\ \text{sem: sg} \\ \text{fnc: peel} \\ \text{prn: 33} \end{bmatrix}
\begin{bmatrix} \text{verb: peel} \\ \text{cat: decl} \\ \text{sem: past} \\ \text{arg: Jim apple} \\ \text{prn: 33} \end{bmatrix}
\begin{bmatrix} \text{noun: Bill} \\ \text{cat: nm} \\ \text{sem: sg} \\ \text{fnc: eat} \\ \text{prn: 33} \end{bmatrix}
\begin{bmatrix} \text{verb: eat} \\ \text{cat: decl} \\ \text{sem: past} \\ \text{arg: Bill apple} \\ \text{prn: 33} \end{bmatrix}
\begin{bmatrix} \text{noun: apple} \\ \text{cat: snp} \\ \text{sem: def sg} \\ \text{fnc: buy} \\ \text{peel} \\ \text{eat} \\ \text{prn: 33} \end{bmatrix}
$$

Here the shared item is the object *apple*, which lists the three verbs in its **fnc** slot, just as the three verbs list **apple** in the second position of their **arg** slot.

While subject, verb, and object gapping are sentential in nature (though intrapropositional), noun gapping applies to the adnominal modifiers inside a phrasal noun,[15] as in the following example:

9.6.4 DBS GRAPH ANALYSIS OF NOUN GAPPING
Bob ate the red #, the green #, and the blue berries.

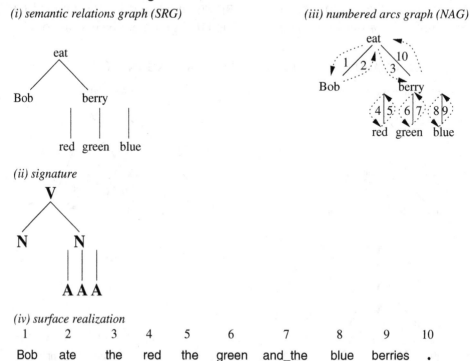

(i) semantic relations graph (SRG)

(iii) numbered arcs graph (NAG)

(ii) signature

(iv) surface realization

1	2	3	4	5	6	7	8	9	10
Bob	ate	the	red	the	green	and_the	blue	berries	.

---

[15] Less than 0.02% of the BNC sentences show this construction of noun gapping.

As in object gapping, the noun gapping parts precede the shared filler. The degree sequence is 321111. The node represented by the noun **berry** is shared by the adnominally used adjectives **red**, **green**, and **blue**, as indicated by the three parallel vertical lines. The surface realization shows no empty traversals, one multiple realization, and no multiple visits. The proplet representation is as follows:

9.6.5 PROPLET REPRESENTATION OF NOUN GAPPING

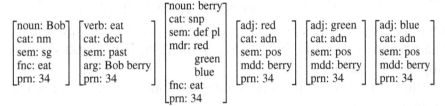

While the noun gapping construction has a *distributive* interpretation in the sense that each berry is either red, green, or blue, the corresponding construction with an adnominal coordination has a *collective* interpretation in the sense that each berry may have all three colors at the same time.

9.6.6 DBS GRAPH ANALYSIS OF ADNOMINAL COORDINATION
Bob ate the red, green, and blue berries.

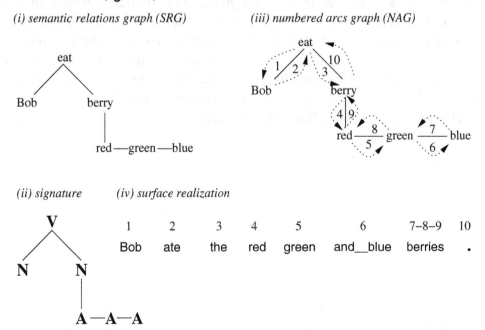

The degree sequence is 222211. Compare the following proplet representation of adnominal coordination with that of noun gapping in 9.6.5:

9.6.7 PROPLET REPRESENTATION OF ADNOMINAL COORDINATION

$$
\begin{bmatrix} \text{noun: Bob} \\ \text{cat: nm} \\ \text{sem: sg} \\ \text{fnc: eat} \\ \text{prn: 35} \end{bmatrix}
\begin{bmatrix} \text{verb: eat} \\ \text{cat: decl} \\ \text{sem: past} \\ \text{arg: Bob berry} \\ \text{prn: 35} \end{bmatrix}
\begin{bmatrix} \text{noun: berry} \\ \text{cat: pnp} \\ \text{sem: def pl} \\ \text{mdr: red} \\ \text{fnc: eat} \\ \text{prn: 35} \end{bmatrix}
\begin{bmatrix} \text{adj: red} \\ \text{cat: adn} \\ \text{sem: pos} \\ \text{mdd: berry} \\ \text{nc: green} \\ \text{pc:} \\ \text{prn: 35} \end{bmatrix}
\begin{bmatrix} \text{adj: green} \\ \text{cat: adn} \\ \text{sem: pos} \\ \text{mdr:} \\ \text{nc: blue} \\ \text{pc: red} \\ \text{prn: 35} \end{bmatrix}
\begin{bmatrix} \text{adj: blue} \\ \text{cat: adn} \\ \text{sem: pos} \\ \text{mdr:} \\ \text{nc:} \\ \text{pc: green} \\ \text{prn: 35} \end{bmatrix}
$$

The connections between the adnominal modifiers and their noun are run via the nc and pc slots of the adjectives – and not via the mdr and mdd slots, as in the gapping construction 9.6.5.

In conclusion consider the following example combining object gapping and noun gapping:

9.6.8 COMBINATION OF OBJECT GAPPING AND NOUN GAPPING

Bob bought #, Jim peeled #, and Bill ate the red #, the green #, and the yellow apple.

In summary, the content corresponding to gapping is an intrapropositional functor-argument, but the surface has a grammatical similarity of function and a superficial resemblance of form with coordination. On the one hand, these particular examples of a discrepancy[16] between a content and the grammatical surface structure may be found in many natural languages, with strong native speaker intuition. On the other hand, empirical investigations of the BNC and the LIMAS corpus have shown that the different gapping constructions are rarely used, at least in English and in German.

---

[16] Another example of this kind is the bare infinitive analyzed in 8.6.2.

10. Computing Perspective in Dialogue

A computational reconstruction of dialogue requires not only an analysis of language expressions, but of *utterances*. In DBS, an utterance is defined as (i) a propositional content and (ii) a cluster of four pointers called STAR.[1] The STAR serves to anchor a content to the interpretation's parameter values of Space, Time, Agent, and Recipient. The STAR also provides the referents for a certain kind of sign in natural language, namely the *indexical*.

Three different STARs are used to code three different perspectives on content: (i) the STAR-0 for coding the agent's perspective on nonlanguage content resulting from recording current recognition and action, (ii) the STAR-1 for coding the speaker's perspective on content underlying language production, and (iii) the STAR-2 for coding the hearer's perspective on content resulting from language interpretation. It is shown that the computation of these perspectives by means of DBS inferences is completely mechanical.

10.1 Agent's STAR-0 Perspective on Current Content

Indexicals raise the question of whether they should be classified as function or as content words. In DBS, all words with their own reference mechanism, i.e., symbol, indexical, and name, are treated as content words (*Autosemantika*). All words which do not have their own reference mechanism, such as determiners, prepositions, conjunctions, and auxiliaries, are treated as function words (*Synsemantika*).

English, like many other natural languages, has five fundamental indexical pointers which are used to refer to (1) the speaker, as in I and we, (2) the hearer, as in you, (3) an agent or object other than the speaker and the hearer, as in he, she, it, they, and this, (4) the present moment, as in now, and (5) the present location, as in here. The referent of an indexical is determined by the *origin* of the sign containing it, represented as the STAR of the sign's utterance.

---

[1] FoCL'99, Sect. 5.3.

However, before a content can be uttered in natural language it must emerge as nonlanguage content in the agent's cognition. Nonlanguage content recording the agent's current situation is anchored to the STAR-0. The following example illustrates an anchored content resulting from recording an activity self-performed by the agent:

10.1.1 ANCHORED NONLANGUAGE CONTENT

I am writing you a letter.$^{\text{STAR}-0}$

In this informal representation, the content is specified in two parts:[2] (i) a sentence of English, used as an aid to represent a nonlanguage content, and (ii) a STAR-0, added as a superscript at the end.[3] The content represented by the English sentence alone is called an *unanchored* content. In DBS, it is coded as the following set of proplets:

10.1.2 CODING UNANCHORED CONTENT AS A PROPLET SET

$$\begin{bmatrix} \text{noun: moi}^4 \\ \text{fnc: write} \\ \text{prn: 659} \end{bmatrix} \begin{bmatrix} \text{verb: write} \\ \text{arg: moi toi letter} \\ \text{prn: 659} \end{bmatrix} \begin{bmatrix} \text{noun: toi} \\ \text{fnc: write} \\ \text{prn: 659} \end{bmatrix} \begin{bmatrix} \text{noun: letter} \\ \text{fnc: write} \\ \text{prn: 659} \end{bmatrix}$$

The values of the associated STAR-0 may be specified as follows:

10.1.3 SPECIFICATION OF A STAR

S = Paris
T = 1930-07-03
A = Jean-Paul Sartre
R = Simone de Beauvoir

While the value of the A attribute is constant for any individual agent, the value of the T attribute is continuously changing with time. The values of the S and R attributes may also change. Therefore, an unanchored content must be connected to the current STAR-0 as soon as the content emerges in the agent's cognition.

The connection between an unanchored content and its STAR-0 is formally established by defining the STAR-0 as a proplet with the same **prn** value as the associated propositional content. The following example complements the unanchored proplet representation 10.1.2 with the STAR-0 anchor 10.1.3:

---

[2] This notation is reminiscent of Montague's (1974) use of "indices." However, while the @,i,j,g index cluster in Montague's PTQ refers to a set-theoretically defined model structure, the STAR refers to the agent-external real world, including the agent viewed from the outside.

[3] The distinction between a content and its anchor may be applied to any system recording its current state. Systems may differ, however, with respect to the attributes and range of values of their anchor.

10.1.4 STAR-0 CONTENT WITH 1ST AND 2ND PERSON INDEXICALS

$$
\begin{bmatrix} \text{noun: moi} \\ \text{fnc: write} \\ \text{prn: 659} \end{bmatrix}
\begin{bmatrix} \text{verb: write} \\ \text{arg: moi toi letter} \\ \text{prn: 659} \end{bmatrix}
\begin{bmatrix} \text{noun: toi} \\ \text{fnc: write} \\ \text{prn: 659} \end{bmatrix}
\begin{bmatrix} \text{noun: letter} \\ \text{fnc: write} \\ \text{prn: 659} \end{bmatrix}
\begin{bmatrix} \text{S: Paris} \\ \text{T: 1930-07-03} \\ \text{A: J.-P. Sartre} \\ \text{R: S. de Beauvoir} \\ \text{prn: 659} \end{bmatrix}
$$

This nonlanguage content constitutes a perspective insofar as J.-P. is looking out towards his current location S, his current moment of time T, himself as an agent in the world A, and Simone as his partner of discourse R. As indexicals, moi and toi[5] are defined to point at the A and R values, respectively, of the STAR-0.

An agent may also register an observation without any self-reference A or involvement of a recipient R, as in the following example.

10.1.5 STAR-0 CONTENT WITHOUT INDEXICALS

$$
\begin{bmatrix} \text{noun: Fido} \\ \text{fnc: bark} \\ \text{prn: 572} \end{bmatrix}
\begin{bmatrix} \text{verb: bark} \\ \text{arg: Fido} \\ \text{prn: 572} \end{bmatrix}
\begin{bmatrix} \text{S: Paris} \\ \text{T: 1930-07-03} \\ \text{A: S. de Beauvoir} \\ \text{prn: 572} \end{bmatrix}
$$

In this anchored nonlanguage content, Simone notes that Fido is barking. Because the content contains no indexical pointing at the R attribute, this feature may be omitted in the STAR-0 specification.

Alternatively to referring to Fido by name, Simone could have used the 3rd person index ça,[6] as in the following example:

10.1.6 STAR-0 CONTENT WITH A 3RD PERSON INDEXICAL

$$
\begin{bmatrix} \text{noun: ça} \\ \text{fnc: bark} \\ \text{prn: 572} \end{bmatrix}
\begin{bmatrix} \text{verb: bark} \\ \text{arg: ça} \\ \text{prn: 572} \end{bmatrix}
\begin{bmatrix} \text{S: Paris} \\ \text{T: 1930-07-03} \\ \text{A: S. de Beauvoir} \\ \text{3rd: Fido} \\ \text{prn: 572} \end{bmatrix}
$$

In avionics and air traffic control, for example, the anchor would use attributes like ground location, altitude, speed, and direction (*situation awareness*, Endsley et al. 2000, 2003) rather than space, time, author, and recipient.

[4] We are using moi and toi instead of I and you because I is not sufficiently distinctive typographically.

[5] In real life, J.-P. Sartre and S. de Beauvoir addressed each other with vous. This relates to the register of honorifics, which is highly grammaticalized in Korean and Japanese, for example. As a socially important aspect of perspective, the register of honorifics (politeness) must be integrated into the R value of the STAR.

[6] In the FoCL'99 definition of the STAR, the attribute 3rd providing the referent for the third person pointer ça was not included because it does not participate directly in anchoring a sign's content. It is necessary, however, for the interpretation of indexicals like this and indexical (i.e., non-coreferential) uses of he, she, it, etc. Rather than changing the terminology from STAR to STAR3rd, we will continue with the simpler term STAR, but use the attribute 3rd and the value ça for third person indexicals when needed (e.g., 10.1.6).

The indexical **ça** is defined to point at the value of the **3rd** attribute of the STAR-0. In contradistinction to 10.1.5, the reference to Fido is not by name here, but by indexical.

For STAR-0 contents, the attributes **R** and **3rd** are optional in that their values need only be defined if the content contains **toi** and **ça** pointers, respectively. For STAR-1 contents (speak mode) and STAR-2 contents (hear mode), a value for the **R** attribute is obligatory; in small children, the **R** value may be viewed as fixed to the mother.

In summary, values for **S**, **T**, and **A** attributes in the STAR are obligatory for all anchored contents. Thereby a rough idea of the location, of the date and the time of day, and of oneself will usually suffice for all practical purposes. Loss of the **STA** parameter values for any length of time, however, results in complete disorientation of the agent.

10.2 Speaker's STAR-1 Perspective on Stored Content

The next day (T: 1930-07-04), Simone and J.-P. meet in the Café de Flore on the Left Bank. To get the conversation going, J.-P. begins a *statement dialogue* by uttering the following declarative sentence to her (in French):

10.2.1 STAR-1 EXPRESSION WITH 1ST AND 2ND PERSON INDEXICALS

I wrote you a letter yesterday.$^{\text{STAR}-1}$

At this point, J.-P. is dealing with two STARs, (i) the STAR-0 defined in 10.1.3 and used in the anchored content 10.1.4 and (ii) the STAR-1 for the utterance indicated in 10.2.1. The **S**, **A**, and **R** values of these two STARs in J.-P.'s mind happen to be the same,[7] but the **T** values differ. This difference constitutes a second kind of perspective: J.-P. is looking back onto a content created in his mind in the recent past. The English surface of 10.2.1 reflects this perspective by means of (i) the past tense form of the finite verb and (ii) the adverbial modifier **yesterday.**

At the level of content, the automatic coding of the speaker perspective is based on the following DBS inference, called STAR-1.1. It takes an anchored STAR-0 and the speaker's current STAR-1 content like 10.1.4 as input:

10.2.2 STAR-1.1 INFERENCE FOR TEMPORAL SPECIFICATION

$$
\begin{bmatrix} \text{verb: } \alpha \\ \text{prn: K} \end{bmatrix}
\begin{bmatrix} \text{S: L} \\ \text{T: D} \\ \text{A: N} \\ \text{R: O} \\ \text{prn: K} \end{bmatrix}
\begin{bmatrix} \text{S: L}' \\ \text{T: D}' \\ \text{A: N} \\ \text{R: O}' \\ \text{prn: K+M} \end{bmatrix}
\Rightarrow
\begin{bmatrix} \text{verb: } \alpha \\ \text{sem: } \beta \\ \text{mdr: } \gamma \\ \text{prn: K+M} \end{bmatrix}
\begin{bmatrix} \text{adj: } \gamma \\ \text{mdd: } \alpha \\ \text{prn: K+M} \end{bmatrix}
\begin{bmatrix} \text{S: L}' \\ \text{T: D}' \\ \text{A: N} \\ \text{R: O}' \\ \text{prn: K+M} \end{bmatrix}
$$

If $D < D'$, then $\beta = \textsf{past}$, and if D diff $D' = 1$ day, then $\gamma = \textsf{yesterday}$; and similarly for all the other possible temporal relations between a STAR-0 and a STAR-1 differing in their T value.

In the input schema, an anchored STAR-0 content is represented by patterns for a verb and for a STAR-0 which share the **prn** variable K (first two proplets). This anchored STAR-0 content is reanchored to the speaker's current STAR-1 with the **prn** value **K+M** (third proplet); its **S**, **T**, and **R** values L′, D′, and O′, respectively, are potentially different from those of the STAR-0 (though the A value N must be the same).

The output is represented by a modified pattern for the verb, an additional proplet pattern for an (optional) adverbial modifier, and the STAR-1 pattern. The restrictions on the variables β and γ are used to control the tense and the temporal adverbial specified in the output schema of the inference.

Applying the inference STAR-1.1 to (i) J.-P.'s current STAR-1 and (ii) the STAR-0 content 10.1.4 results in the following STAR-1 content (speak mode):

10.2.3 STAR-1 CONTENT *Moi* wrote *toi* a letter yesterday.

$$
\begin{bmatrix} \text{noun: moi} \\ \text{fnc: write} \\ \text{prn: 659+7} \end{bmatrix}
\begin{bmatrix} \text{verb: write} \\ \text{arg: moi toi letter} \\ \text{sem: past} \\ \text{mdr: yesterday} \\ \text{prn: 659+7} \end{bmatrix}
\begin{bmatrix} \text{noun: toi} \\ \text{fnc: write} \\ \text{prn: 659+7} \end{bmatrix}
\begin{bmatrix} \text{noun: letter} \\ \text{fnc: write} \\ \text{prn: 659+7} \end{bmatrix}
\begin{bmatrix} \text{adj: yesterday} \\ \text{mdd: write} \\ \text{prn: 659+7} \end{bmatrix}
\begin{bmatrix} \text{S: Paris} \\ \text{T: 1930-07-04} \\ \text{A: J.-P. Sartre} \\ \text{R: S. de Beauvoir} \\ \text{prn: 659+7} \end{bmatrix}
$$

Compared to 10.1.4, the **sem** attribute of the verb has received the value **past**, and the adverbial modifier **yesterday** has been added as a proplet connected to the verb. Also, the new **prn** value **659+7** has been assigned by the parser not just to the verb and adj proplets matched by the output schema of the STAR-1.1 inference, but to all proplets of the resulting content.

The output of the inference is written to the *now front* of the agent's Word Bank. Thus, the original content and its STAR-0 are not overwritten. They may be retrieved by using the first part of the new content's **prn** value, here **659**.

This is in concord with the content-addressable database of a Word Bank, in which content is written once and never changed (Sect. 4.1). The antecedent of a DBS inference may only *read* stored content, while the consequent may only *write* to the now-front of the Word Bank – as reflected by the **prn** values of the DBS inferences 10.2.2 and 10.2.4.

In addition to the temporal respecification provided by the STAR-1.1 inference there must be a STAR-1.2 inference for a possible respecification of the S

---

[7] For simplicity, the difference in location between J.-P.'s apartment and the Café de Flore within Paris is not reflected by the STARs' S values. Also, whole days are used for the T values. This *soft* treatment of the spatio-temporal coordinates is more appropriate for modeling a natural agent's cognition than are the unnaturally precise values provided by the natural sciences, though they may be used if needed.

value and a STAR-1.3 inference for a possible respecification of the R value. For example, if J.-P. were to meet Juliette instead of Simone at the Café de Flore, the R value of his STAR-1 would be Juliette and the content of 10.1.4 would be realized as **I wrote Simone a letter yesterday**. The STAR-1.3 inference is defined as follows:

10.2.4 STAR-1.3 INFERENCE FOR SPECIFICATION OF RECIPIENT

$$
\begin{bmatrix} \text{verb: } \alpha \\ \text{arg: } \{X \text{ toi}\} \\ \text{prn: K} \end{bmatrix}
\begin{bmatrix} \text{noun: toi} \\ \text{fnc: } \alpha \\ \text{prn: K} \end{bmatrix}
\begin{bmatrix} \text{S: L} \\ \text{T: D} \\ \text{A: N} \\ \text{R: O} \\ \text{prn: K} \end{bmatrix}
\begin{bmatrix} \text{S: L}' \\ \text{T: D}' \\ \text{A: N} \\ \text{R: O}' \\ \text{prn: K+M} \end{bmatrix}
\Rightarrow
\begin{bmatrix} \text{verb: } \alpha \\ \text{arg: } \{X \text{ O}\} \\ \text{prn: K+M} \end{bmatrix}
\begin{bmatrix} \text{noun: O} \\ \text{fnc: } \alpha \\ \text{prn: K+M} \end{bmatrix}
\begin{bmatrix} \text{S: L}' \\ \text{T: D}' \\ \text{A: N} \\ \text{R: O}' \\ \text{prn: K+M} \end{bmatrix}
$$

The first three pattern proplets of the antecedent match the STAR-0 content 10.1.4. The consequent (output pattern) replaces toi by the R value O of the STAR-0 and assigns the prn value of the STAR-1. The [arg: {X toi}] specification in the input verb pattern is intended to match toi as subject or object. In this way, the inference may result in contents like **I wrote Simone a letter** as well as **Simone wrote me a letter**.

Applying STAR-1.1 and STAR-1.3 to 10.1.4 results in the following content:

10.2.5 STAR-1 CONTENT *Moi* wrote Simone a letter yesterday.

$$
\begin{bmatrix} \text{noun: moi} \\ \text{fnc: write} \\ \text{prn: 659+7} \end{bmatrix}
\begin{bmatrix} \text{verb: write} \\ \text{arg: moi Simone letter} \\ \text{sem: past} \\ \text{mdr: yesterday} \\ \text{prn: 659+7} \end{bmatrix}
\begin{bmatrix} \text{noun: Simone} \\ \text{fnc: write} \\ \text{prn: 659+7} \end{bmatrix}
\begin{bmatrix} \text{noun: letter} \\ \text{fnc: write} \\ \text{prn: 659+7} \end{bmatrix}
\begin{bmatrix} \text{adj: yesterday} \\ \text{mdd: write} \\ \text{prn: 659+7} \end{bmatrix}
\begin{bmatrix} \text{S: Paris} \\ \text{T: 1930-07-04} \\ \text{A: J.-P. Sartre} \\ \text{R: Juliette} \\ \text{prn: 659+7} \end{bmatrix}
$$

The derivation of STAR-1 contents does not interfere with the cycle of natural language communication, because it applies after the activation of content for language production by the think mode and prior to the LA-speak derivation of a corresponding surface.

10.3 Hearer's STAR-2 Perspective on Language Content

When Simone hears the utterance 10.2.1, she does a standard time-linear LA-hear derivation, resulting in the following set of proplets:

10.3.1 RESULT OF ANALYZING 10.2.1 IN THE HEAR MODE

$$
\begin{bmatrix} \text{noun: moi} \\ \text{fnc: write} \\ \text{prn: 623} \end{bmatrix}
\begin{bmatrix} \text{verb: write} \\ \text{arg: moi toi letter} \\ \text{sem: past} \\ \text{mdr: yesterday} \\ \text{prn: 623} \end{bmatrix}
\begin{bmatrix} \text{noun: toi} \\ \text{fnc: write} \\ \text{prn: 623} \end{bmatrix}
\begin{bmatrix} \text{noun: letter} \\ \text{fnc: write} \\ \text{prn: 623} \end{bmatrix}
\begin{bmatrix} \text{adj: yesterday} \\ \text{mdd: write} \\ \text{prn: 623} \end{bmatrix}
$$

As the result of a strictly surface compositional reconstruction of the speaker's literal meaning, the content 10.3.1 represents the perspective of the speaker J.-P. – except for the prn value 623, which equals that of Simone's current STAR.

In the hear mode, the following main perspectives[8] on incoming speak mode content must be distinguished between:

10.3.2 MAIN HEAR MODE PERSPECTIVES ON LANGUAGE CONTENT

1. The perspective of the hearer as the partner in face-to-face communication.
2. The perspective of someone overhearing a conversation between others.
3. The reader's perspective onto the content of a written text (Chap. 11).

Given that Simone and J.-P. are partners in face-to-face communication, the correct way for Simone to convert J.-P.'s speak mode perspective is from

I wrote you a letter yesterday.$^{\text{STAR}-1}$

to her own hear mode perspective as

You wrote me a letter yesterday.$^{\text{STAR}-2}$

This automatic conversion is based on the STAR-2.1 inference (see 10.3.3 below), which takes a content like 10.3.1 and a STAR-1 as input. The STAR-1 is *attributed* by the hearer to the speaker. In face-to-face communication, this is easy because the speaker's STAR-1 and the hearer's STAR-2 correspond in that their S and T values are the same, and their A and R values are reversed. The DBS inference STAR-2.1 is defined as follows:

10.3.3 STAR-2.1 INFERENCE FOR DERIVING HEARER PERSPECTIVE

$$
\begin{bmatrix} \text{noun: moi} \\ \text{fnc: } \alpha \\ \text{prn: K} \end{bmatrix}
\begin{bmatrix} \text{verb: } \alpha \\ \text{arg: \{X moi toi\}} \\ \text{prn: K} \end{bmatrix}
\begin{bmatrix} \text{noun: toi} \\ \text{fnc: } \alpha \\ \text{prn: K} \end{bmatrix}
\begin{bmatrix} \text{S: L} \\ \text{T: D} \\ \text{A: N} \\ \text{R: O} \\ \text{prn: K} \end{bmatrix} \Rightarrow
$$

$$
\begin{bmatrix} \text{noun: toi} \\ \text{fnc: } \alpha \\ \text{prn: K} \end{bmatrix}
\begin{bmatrix} \text{verb: } \alpha \\ \text{arg: \{X toi moi\}} \\ \text{prn: K} \end{bmatrix}
\begin{bmatrix} \text{noun: moi} \\ \text{fnc: } \alpha \\ \text{prn: K} \end{bmatrix}
\begin{bmatrix} \text{S: L} \\ \text{T: D} \\ \text{A: O} \\ \text{R: N} \\ \text{prn: K} \end{bmatrix}
$$

In the output, the speaker's STAR-1 perspective is revised into the hearer's current STAR-2 perspective by swapping the A and R values N and O, and

---

[8] Special cases are phone conversations which require the hearer to recompute the speaker's S(pace) value; when talking across time zones, the speaker's T(ime) value must be recomputed as well.

keeping the S, T, and **prn** values. The notation {X moi toi} is intended to indicate that moi, toi, and X (for other category segments) will match content conterparts in any order.

If we assume that the STAR-2 used by Simone is [S: Paris], [T: 1930-07-04], [A: Simone], and [R: J.-P.], then the application of the inference STAR-2.1 to content 10.3.1 results in the following STAR-2 content:

10.3.4 STAR-2 CONTENT *Toi* wrote *moi* a letter yesterday.

$$
\begin{bmatrix} \text{noun: toi} \\ \text{fnc: write} \\ \text{prn: 623} \end{bmatrix}
\begin{bmatrix} \text{verb: write} \\ \text{arg: toi moi letter} \\ \text{sem: past} \\ \text{mdr: yesterday} \\ \text{prn: 623} \end{bmatrix}
\begin{bmatrix} \text{noun: moi} \\ \text{fnc: write} \\ \text{prn: 623} \end{bmatrix}
\begin{bmatrix} \text{noun: letter} \\ \text{fnc: write} \\ \text{prn: 623} \end{bmatrix}
\begin{bmatrix} \text{adj: yesterday} \\ \text{mdd: write} \\ \text{prn: 623} \end{bmatrix}
\begin{bmatrix} \text{S: Paris} \\ \text{T: 1930-07-04} \\ \text{A: Simone de B.} \\ \text{R: J.-P. Sartre} \\ \text{prn: 623} \end{bmatrix}
$$

Here, **toi** is pointing at the **R** value J.-P., **moi** is pointing at the **A** value Simone, **yesterday** is pointing at the **T** value 1930-07-04, and the **sem** attribute of the verb has the value **past** from the hear mode analysis of the surface.

As an example without any indexicals consider Simone producing the following utterance addressed to J.-P., using her STAR-0 content 10.1.5:

10.3.5 STAR-1 CONTENT WITHOUT INDEXICALS

Fido barked.[STAR−1]

For the hearer J.-P., the interpretation of this content requires adjusting Simone's STAR-1 speak mode perspective to J.-P.'s STAR-2 hear mode perspective. This is based on the following STAR-2.2 inference:

10.3.6 STAR-2.2 INFERENCE FOR CONTENT WITHOUT INDEXICALS

$$
\begin{bmatrix} \text{verb: } \alpha \\ \text{prn: K} \end{bmatrix}
\begin{bmatrix} \text{S: L} \\ \text{T: D} \\ \text{A: N} \\ \text{R: O} \\ \text{prn: K} \end{bmatrix}
\Rightarrow
\begin{bmatrix} \text{verb: } \alpha \\ \text{prn: K} \end{bmatrix}
\begin{bmatrix} \text{S: L} \\ \text{T: D} \\ \text{A: O} \\ \text{R: N} \\ \text{prn: K} \end{bmatrix}
$$

J.-P. assigns Simone's STAR-1 to the input by swapping the A and R values of his STAR-2, coding J.-P.'s perspective, with the following result:

10.3.7 STAR-2 CONTENT **Fido barked.**

$$
\begin{bmatrix} \text{noun: Fido} \\ \text{fnc: bark} \\ \text{prn: 572} \end{bmatrix}
\begin{bmatrix} \text{verb: bark} \\ \text{arg: Fido} \\ \text{sem: past} \\ \text{prn: 572} \end{bmatrix}
\begin{bmatrix} \text{S: Paris} \\ \text{T: 1930-07-03} \\ \text{A: J.-P. Sartre} \\ \text{R: Simone de B.} \\ \text{prn: 572} \end{bmatrix}
$$

The properties of STAR-2 contents resulting from STAR-1 contents transmitted in face-to-face communication may be summarized as follows:

10.3.8 OPERATIONS OF STAR-2 INFERENCES

1. The S value of the STAR-1 in the input (matching the antecedent) equals the S value of the STAR-2 in the output (derived by the consequent).

2. The T value of the STAR-1 in the input equals the T value of the STAR-2 in the output.

3. The A value of the STAR-1 in the input equals the R value of the STAR-2 in the output.

4. The R value of the STAR-1 in the input equals the A value of the STAR-2 in the output.

5. The prn value of the input equals the prn value of the output.

These properties hold specifically for STAR-2 contents. For example, in STAR-1 contents the author equals the A value, and not the R value.

The derivation of STAR-2 contents does not interfere with the cycle of natural language communication (Sects. 3.3, 3.4) because it applies after the hear mode derivation and before storage in the Word Bank.

10.4 Dialogue with a WH Question and Its Answer

The statement dialogue analyzed in Sects. 10.1–10.3 consists of the speaker producing and the hearer interpreting a declarative[9] sentence. A question-answer dialogue, in contrast, is based on (1) the questioner producing an interrogative, (2) the answerer interpreting the interrogative, (3) the answerer producing an answer, and (4) the questioner interpreting the answer.

Preceding these four steps, however, there is the emergence of the question *content*. For example, having digested J.-P.'s remark 10.2.1, Simone searches her recent memory for connected **letter**, **write**, and **J.-P.** proplets, and realizes that she has not yet received the letter.[10] This creates a certain kind of imbalance in her mind, commonly known as curiosity. As a means to regain her equilibrium, the following question content emerges in Simone's mind:

---

[9] The sentential mood associated with statement dialogues is *declarative*, just as the sentential mood associated with question dialogues is *interrogative*, while the *imperative* mood is associated with requests. However, the sentential moods are often used indirectly. For example, the yes/no interrogative **Could you pass the salt?** is normally used as as a request rather than a question. Similarly, the statement **I demand a signed report now.** (indirect request) is answered in the movie **Bullit** (1968) with the politeness formula **Excuse me.** with the speaker walking away (indirect way of saying **no**). In this chapter, statement, question, and request dialogues are illustrated with literally used expressions in the declarative, interrogative, and imperative moods, respectively.

[10] Despite the much praised postal service by pneumatic delivery in Paris at the time (Beauvoir 1960).

10.4.1 NONLANGUAGE CONTENT IN THE INTERROGATIVE MOOD

What did you write?[STAR-0]

In analogy to 10.1.1–10.1.4, the content and its anchor may be represented here as the following set of proplets:

10.4.2 ANCHORED STAR-0 CONTENT OF WH INTERROGATIVE

$$
\begin{bmatrix} \text{noun: toi} \\ \text{fnc: write} \\ \text{prn: 625} \end{bmatrix}
\begin{bmatrix} \text{verb: write} \\ \text{cat: interrog} \\ \text{sem: past} \\ \text{arg: toi what} \\ \text{prn: 625} \end{bmatrix}
\begin{bmatrix} \text{noun: what} \\ \text{fnc: write} \\ \text{prn: 625} \end{bmatrix}
\begin{bmatrix} \text{S: Paris} \\ \text{T: 1930-07-04} \\ \text{A: Simone de B.} \\ \text{R: J.-P. Sartre} \\ \text{prn: 625} \end{bmatrix}
$$

As in 10.1.4, the connection between the STAR and the content is established by a common **prn** value, here **625**. The indexical **toi** points at the R value of the STAR-0.

Given that there is no significant time difference between the formation of the content and its use for language production, there is no need to derive a separate speaker perspective on the content (in contrast to the transition from 10.1.1 to 10.2.1). Instead, Simone proceeds to realize the surface in 10.4.1 by using (i) 10.4.2 as a STAR-1 content and (ii) the following DBS graph analysis:

10.4.3 QUESTIONER AS SPEAKER: DBS GRAPH ANALYSIS OF 10.4.1

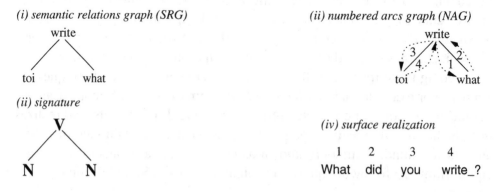

(i) semantic relations graph (SRG)

(ii) numbered arcs graph (NAG)

(ii) signature

(iv) surface realization

1	2	3	4
What	did	you	write_?

As shown in Sect. 7.4, a NAG and a proplet representation of a content (here 10.4.2) are sufficient to quasi-automatically derive an LA-speak grammar which realizes a corresponding surface.

Now it is J.-P.'s turn to interpret the incoming surface (presented in English):

10.4.4 Answerer as Hearer Parsing 10.4.1

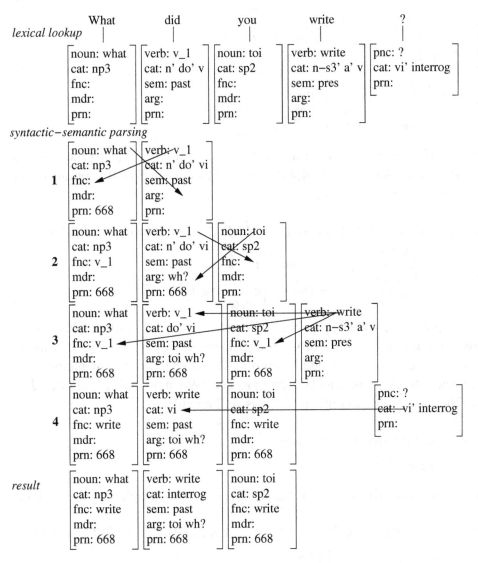

The result of this strictly time-linear, surface compositional derivation is a content which represents the perspective of the speaker Simone – except for the prn value, here **668**, which is assigned by the hearer J.-P.

For understanding, the answerer J.-P. must change Simone's speak mode perspective into his own hear mode perspective by transforming the content *Toi wrote what?*[11] of Simone's question into *Moi wrote what?*, based on the following STAR-2.3 inference:

---

[11] The result of a hear mode derivation is a *set* of proplets, i.e., order-free. In other words, the sets *Toi wrote what?* and *What wrote toi?* are equivalent.

10.4.5 STAR-2.3 INFERENCE FOR DERIVING HEARER PERSPECTIVE

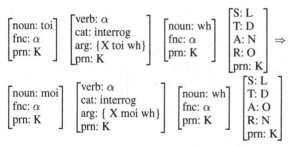

$$
\begin{bmatrix} \text{noun: toi} \\ \text{fnc: } \alpha \\ \text{prn: K} \end{bmatrix}
\begin{bmatrix} \text{verb: } \alpha \\ \text{cat: interrog} \\ \text{arg: } \{X \text{ toi wh}\} \\ \text{prn: K} \end{bmatrix}
\begin{bmatrix} \text{noun: wh} \\ \text{fnc: } \alpha \\ \text{prn: K} \end{bmatrix}
\begin{bmatrix} \text{S: L} \\ \text{T: D} \\ \text{A: N} \\ \text{R: O} \\ \text{prn: K} \end{bmatrix} \Rightarrow
$$

$$
\begin{bmatrix} \text{noun: moi} \\ \text{fnc: } \alpha \\ \text{prn: K} \end{bmatrix}
\begin{bmatrix} \text{verb: } \alpha \\ \text{cat: interrog} \\ \text{arg: } \{X \text{ moi wh}\} \\ \text{prn: K} \end{bmatrix}
\begin{bmatrix} \text{noun: wh} \\ \text{fnc: } \alpha \\ \text{prn: K} \end{bmatrix}
\begin{bmatrix} \text{S: L} \\ \text{T: D} \\ \text{A: O} \\ \text{R: N} \\ \text{prn: K} \end{bmatrix}
$$

This STAR-2 inference complies with 10.3.8. Applying it to (i) the result of the hear mode derivation 10.4.4 and (ii) J.-P.'s current STAR-2 produces the following anchored content:

10.4.6 RESULT OF APPLYING THE STAR-2.3 INFERENCE TO 10.4.4

$$
\begin{bmatrix} \text{noun: moi} \\ \text{fnc: write} \\ \text{prn: 668} \end{bmatrix}
\begin{bmatrix} \text{verb: write} \\ \text{cat: interrog} \\ \text{sem: past} \\ \text{arg: moi wh} \\ \text{prn: 668} \end{bmatrix}
\begin{bmatrix} \text{noun: what} \\ \text{fnc: write} \\ \text{prn: 668} \end{bmatrix}
\begin{bmatrix} \text{S: Paris} \\ \text{T: 1930-07-04} \\ \text{A: J.-P. Sartre} \\ \text{R: Simone de B.} \\ \text{prn: 668} \end{bmatrix}
$$

At this point, the answerer J.-P. understands Simone's question. This has the effect of passing Simone's original imbalance successfully on to J.-P. as the hearer. To reestablish his equilibrium, J.-P. searches his recent memory for connected **letter**, **write**, and **Simone** proplets. When he finds the answer, J.-P. uses the speak mode to reply as follows (in French):

10.4.7 ANSWERER AS SPEAKER

A little poem.[STAR−1]

The content underlying this answer has the following proplet representation:

10.4.8 ANSWER TO A WH QUESTION AS A SET OF STAR-0 PROPLETS

$$
\begin{bmatrix} \text{noun: poem} \\ \text{sem: indef sg} \\ \text{mdr: little} \\ \text{prn: 655} \end{bmatrix}
\begin{bmatrix} \text{adj: little} \\ \text{mdd: poem} \\ \text{prn: 655} \end{bmatrix}
\begin{bmatrix} \text{S: Paris} \\ \text{T: 1930-07-03} \\ \text{A: J.-P. Sartre} \\ \text{R: Simone de B.} \\ \text{prn: 655} \end{bmatrix}
$$

A pertinent answer must *precede* the question at the level of content in the mind of the answerer.[12] It is only when the answer is realized in language that it follows the question, for answerer and questioner alike.

The final turn of a question-answer sequence is the questioner in the hear mode. In our example, Simone as the hearer parses J.-P.'s answer as follows:

---

[12] As reflected by the prn values **668** of the question 10.4.6 and **655** of the answer 10.4.8.

10.4.9 QUESTIONER AS HEARER PARSING 10.4.7

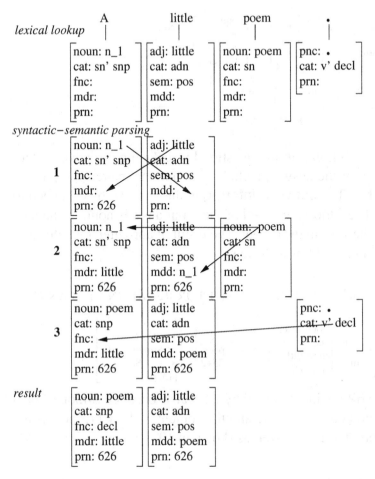

On the one hand, the answer to a WH question is declarative in nature, as indicated by the full stop. On the other hand, it is missing a verb. To characterize the result of the above derivation as a WH answer, we propose copying the value **decl** of the full stop into the **fnc** slot of the noun, and using the [fnc: decl] feature as a marker for WH answers. This characterizes WH answers uniquely and allows easy definition of a pattern proplet for their retrieval.

While balance is reestablished for the answerer when uttering the answer (10.4.7), the questioner must not only interpret the answer, as in 10.4.9, but combine the WH question and the answer into one declarative content. Derived by the following STAR-2 inference, the resulting content has a new **prn** value:

10.4.10 STAR-2.4 CONNECTING WH INTERROGATIVE WITH ANSWER

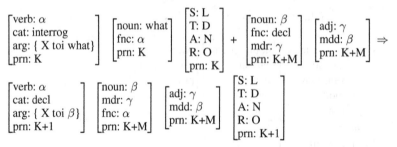

The inference fills the wh gap in the question by replacing the *what* pattern proplet with the noun of the answer, including an optional pattern proplet for an adnominal modifier. The cat value interrog in the input verb is changed to decl in the output. The func value decl of the input answer noun is changed to the core value of the verb in the output. The result of applying this inference to the inputs 10.4.2 and 10.4.8 is as follows:

10.4.11 QUESTIONER'S STAR-2 CONTENT FOR REGAINING BALANCE

$$
\begin{bmatrix} \text{noun: toi} \\ \text{fnc: write} \\ \text{prn: 625+2} \end{bmatrix}
\begin{bmatrix} \text{verb: write} \\ \text{cat: decl} \\ \text{sem: past} \\ \text{arg: toi poem} \\ \text{prn: 625+2} \end{bmatrix}
\begin{bmatrix} \text{noun: poem} \\ \text{fnc: write} \\ \text{mdr: little} \\ \text{prn: 625+2} \end{bmatrix}
\begin{bmatrix} \text{adj: little} \\ \text{mdd: poem} \\ \text{prn: 625+2} \end{bmatrix}
\begin{bmatrix} \text{S: Paris} \\ \text{T: 1930-07-04} \\ \text{A: Simone de B.} \\ \text{R: J.-P. Sartre} \\ \text{prn: 625+2} \end{bmatrix}
$$

The input prn value 625 is incremented by 2 because between Simone's production of the question content 10.4.2 and the content of 10.4.11, there is Simone's interpretation of J.-P.'s answer, as shown in 10.4.9.

10.5 Dialogue with a Yes/No Question and Its Answer

WH questions may request noun values, as in What did you write?, or adjective values, as in Why did you go? (cause), Where did you stay? (location), How did you sleep? (manner), and When did you leave? (time). They all conform to the time-linear structure of question-answer sequences (10.6.7), and their DBS analysis closely resembles that shown in the previous section for a WH question with a noun answer.

Unlike WH questions, yes/no questions request a choice between only two values, namely yes and no, in their various guises. For example, after J.-P.'s answer A little poem., Simone's curiosity is not yet completely satisfied. As a potential countermeasure, the following nonlanguage STAR-0 content emerges in her mind:

10.5.1 STAR-0 CONTENT UNDERLYING LANGUAGE COUNTERMEASURE

Is the poem about me?[STAR−0]

As in 10.4.1 and 10.4.2, this content is used by Simone as a STAR-1 content and realized as a surface (in French). From this unanalyzed external surface, presented in English for convenience, the answerer J.-P. derives the following content:

10.5.2 ANSWERER AS HEARER PARSING A YES/NO INTERROGATIVE

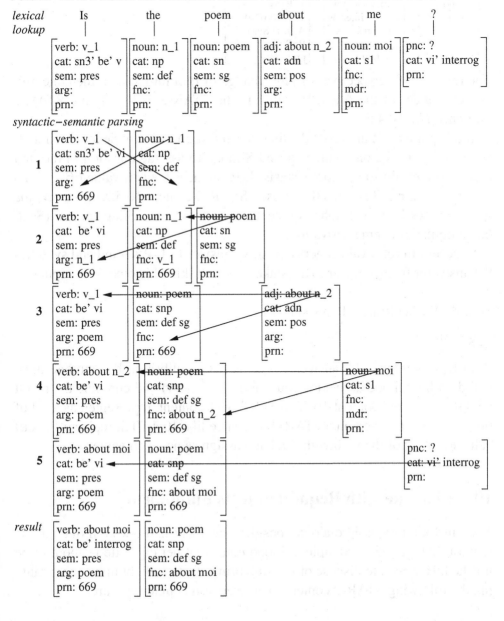

The result codes the perspective of the questioner Simone as speaker, except for the prn value, here 669, which is assigned by the answerer J.-P. as hearer.

Before storage in J.-P.'s Word Bank, Simone's speaker perspective coded as poem is about *moi*? must be revised into the perspective of the answerer as hearer, coded as poem is about *toi*?. The latter is represented as the following set of proplets:

10.5.3 ANSWERER AS HEARER: REVISED PERSPECTIVE OF 10.5.2

$$
\begin{bmatrix} \text{noun: poem} \\ \text{cat: snp} \\ \text{sem: indef sg} \\ \text{fnc: about toi} \\ \text{prn: 669} \end{bmatrix}
\begin{bmatrix} \text{verb: about toi} \\ \text{cat: be' interrog} \\ \text{sem: pres} \\ \text{arg: poem} \\ \text{prn: 669} \end{bmatrix}
\begin{bmatrix} \text{S: Paris} \\ \text{T: 1930-07-04} \\ \text{A: J.-P. Sartre} \\ \text{R: Simone de B.} \\ \text{prn: 669} \end{bmatrix}
$$

The revision of perspective, i.e., the change of moi into toi and the concomitant exchanges of the A and R values in the STAR-2, is based on a STAR-2 inference like 10.4.5.

At this point, J.-P. understands Simone's question and experiences an imbalance similar to the one which caused Simone to ask the question in the first place. To reestablish his homeostasis, J.-P. replaces the prn value 669 with a variable, e.g., K. This turns the revised STAR-2 content 10.5.3 into a schema which allows J.-P. to search his recent memory for a matching content (Sect. 6.5, automatic schema derivation.)

A successful retrieval triggers a positive answer in J.-P.'s mind and switches the answerer from a hearer to a speaker. After stoking his pipe, J.-P. replies:

10.5.4 ANSWERER J.-P. AS SPEAKER

Yes.[STAR-1]

With this utterance, the answerer regains his equilibrium, but Simone must still do a hear mode derivation (questioner as hearer) and combine the result with the anchored content 10.5.1 as a STAR-1 content, represented as a set of proplets. This is based on a STAR-2 inference like 10.4.10; it replaces the cat value interrog of the verb with decl and assigns a new prn value.

10.6 Dialogue with Request and Its Fulfillment

The third kind of basic dialogue besides statement and question dialogue is request dialogue. For example, Simone notes that her cigarette is about to be finished. Her need to dispose of the stub brings forth a slight imbalance, causing the following STAR-0 content to emerge as a countermeasure:

10.6.1 ANCHORED NONLANGUAGE REQUEST CONTENT

(Please)[13] pass the ashtray![STAR−0]

Such a countermeasure may be learned as a one-step inference chain like 5.2.2.
The STAR-0 content 10.6.1 is represented as the following set of proplets:

10.6.2 REQUEST STAR-0 CONTENT AS A SET OF PROPLETS

$$
\begin{bmatrix}
\text{verb: pass} \\
\text{cat: impv} \\
\text{sem: pres} \\
\text{arg: \# ashtray} \\
\text{prn: 630}
\end{bmatrix}
\begin{bmatrix}
\text{noun: ashtray} \\
\text{cat: snp} \\
\text{sem: def sg} \\
\text{fnc: pass} \\
\text{prn: 630}
\end{bmatrix}
\begin{bmatrix}
\text{S: Paris} \\
\text{T: 1930-07-04} \\
\text{A: Simone de B.} \\
\text{R: J.-P. Sartre} \\
\text{prn: 630}
\end{bmatrix}
$$

The requestor equals the **A** value Simone and the requestee equals the **R** value
J.-P. The verb's **cat** value **impv**, for imperative, shows the sentential mood.[14]

After a long last drag on her cigarette, Simone produces the surface in 10.6.1
based on the proplet set 10.6.2 and the following DBS graph structure:

10.6.3 GRAPH STRUCTURE USED BY REQUESTOR AS SPEAKER

Production of the content 10.6.1 as a surface constitutes step 1 of the question-
answer sequence. Given that there is no significant time difference between the
formation of the content and its use for language production, there is no need
to derive a separate speaker perspective. In other words, Simone may reuse the
STAR-0 of 10.6.2 as the STAR-1 of her utterance.

In the next step (i.e., step 2, requestee as the hearer), J.-P. parses the surface
in 10.6.1 as follows:

---

[14] For simplicity, we omit the analysis of **please**. Its pragmatic role is to indicate a polite attitude
towards the hearer, similarly to the use of **vous** instead of **toi**. The syntactic-semantic treatment of
please may either be as an adverbial modifier, or be integrated into a general treatment of *particles*
(Ickler 1994). The widespread natural language phenomenon of particles may require a treatment in
terms of pragmatics, relatively independently of functor-argument and coordination.

[14] The linguistik terms *imperfect* (**imp**), *implies* (**impl**), and *imperative* (**impv**) must be distinguished.

10.6.4 REQUESTEE AS HEARER PARSING Pass the ashtray!

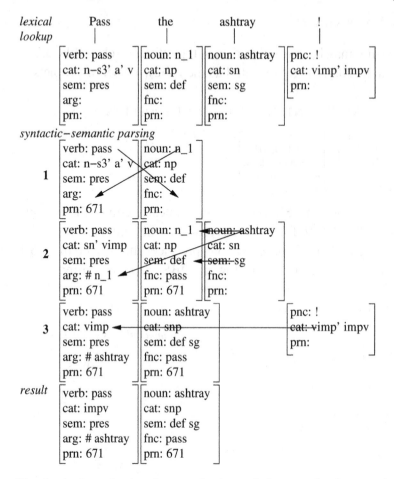

The lexical analysis of **pass** is that of the standard unmarked present tense form of a verb.[15]

The conversion of the lexical **cat** value **n-s3' a' v** to **a' vimp** is accomplished by the LA-hear rule which combines a sentence-initial finite verb with

---

[15] The main sentential moods in natural language are the declarative, the interrogative, and the imperative. These sentential moods must be distinguished from the verbal moods, the indicative, the subjunctive, and the imperative in grammars for English (Sect. 5.6). Thus the term "imperative" has been used for both the sentential and the verbal moods.

In some languages the verbal moods are realized as special forms of the verbal morphology, with dedicated uses in the associated sentential moods. Classical Latin, for example, has separate word forms for the verb in the imperative mood, differentiated by whether the requestee is a singular individual (**curre**, *run* impv. sing.) or a plural group (**currete**, *run* impv. pl.).

English, in contrast, uses the unmarked form of the verb's present tense for constructing the imperative as a sentential mood. Thus, just as there is no separate verbal form for the infinitive (Sect. 8.4), there is no separate verbal form for the imperative. Instead, the imperative as a sentential mood is built as a special syntactic-semantic construction, characterized by word order, intonation or interpunctuation, and the absence of a subject, tense, and verbal mood.

a noun. The result of this hear mode derivation equals the content 10.6.2 of the requestor as speaker, except for the **prn** value and a new STAR-2:

10.6.5 REQUEST STAR-2 CONTENT AS A SET OF PROPLETS

$$
\begin{bmatrix} \text{verb: pass} \\ \text{cat: impv} \\ \text{sem: pres} \\ \text{arg: \# ashtray} \\ \text{prn: 671} \end{bmatrix}
\begin{bmatrix} \text{noun: ashtray} \\ \text{cat: snp} \\ \text{sem: def sg} \\ \text{fnc: pass} \\ \text{prn: 671} \end{bmatrix}
\begin{bmatrix} \text{S: Paris} \\ \text{T: 1930-07-04} \\ \text{A: J.-P. Sartre} \\ \text{R: Simone de B.} \\ \text{prn: 671} \end{bmatrix}
$$

Here the conversion from Simone's perspective as the requestor to J.-P.'s perspective as the requestee is limited to exchanging the **A** and **R** values in the STAR-1 (= STAR-0) of 10.6.2, as compared to the STAR-2 of 10.6.5. This has an effect similar to going from *toi* **pass ashtray** to *moi* **pass ashtray**, though the imperative content has no explicit subject, and consequently no indexical.

At this point, J.-P. understands the request: the slight imbalance experienced by Simone has been successfully passed on to him by means of natural language. As step 3 of a time-linear request sequence, the requestee must take the requested action in order to be cooperative. This may be a nonlanguage action, as responding to **Open the window!**, or a language action, as responding to **Tell me more about your mother!** (Weizenbaum 1965).

In our example, the requestee J.-P. has been asked to perform a nonlanguage action. Therefore, he is looking around for the means to realize his blueprint for action, which may be paraphrased as *Moi* **pass ashtray**. He recognizes the ashtray on the restaurant table (we are talking about the year 1930) and initiates a manipulation sequence which moves the ashtray over to Simone. This results in J.-P. regaining his equilibrium.

The final step 4 is the requestor recognizing the fulfillment of the request by the requestee. In our case, Simone puts out her cigarette in the ashtray. This results in her regaining her balance and closes the sequence.

The main difference between the time-linear sequence of a question and a request dialogue is that steps 3 and 4 in a question dialogue are normally realized as language action, while the corresponding steps in a request dialogue may be either language or nonlanguage action. However, because all essential dialogue operations, i.e., adjustments of perspective and other inferencing, are performed at the level of content, DBS can handle request dialogues just as well as question dialogues.

In summary, this chapter has analyzed dialogue as a sequence of elementary dialogues, using a fictional conversation between Jean-Paul Sartre and Simone de Beauvoir in the Café de Flore on 1930-07-04 as our example:

10.6.6 SEQUENCE OF ELEMENTARY DIALOGUES

J.-P. Sartre:	I wrote you a letter yesterday.	(statement, Sect. 10.1–10.3)
S. de Beauvoir:	What did you write?	(WH question, Sect. 10.4)
J.-P. Sartre:	A little poem.	(WH answer, Sect. 10.4)
S. de Beauvoir:	Is the poem about me?	(Yes/No question, Sect. 10.5)
J.-P. Sartre :	Yes.	(Yes/No answer, Sect. 10.5)
S. de Beauvoir:	(Please) pass the ashtray!	(request, Sect. 10.6)
J.-P. Sartre:	fullfils request	(fullfilment, Sect. 10.6)

Each of the elementary dialogues is a characteristic sequence of turns which consist of the following perspective conversions:

10.6.7 PERSPECTIVE CONVERSIONS AS TIME-LINEAR SEQUENCES

1. **Statement**
   STAR-0: emergence of a nonlanguage content in agent A (Sect. 10.1)
   STAR-1: production of a statement by agent A as the speaker (Sect. 10.2)
   STAR-2: interpretation of statement by agent B as the hearer (Sect. 10.3)

2. **Question Dialogue** (WH (Sect. 10.4) and Yes/No (Sect.10.5) questions)
   STAR-0: emergence of a nonlang. content in agent A as the questioner
   STAR-1: production of a question by agent A as the speaker
   STAR-2: interpretation of the question by agent B as the hearer
   STAR-1: production of an answer by agent B as the speaker
   STAR-2: interpretation of the answer by agent A as the hearer

3. **Request Dialogue** (Sect. 10.6)
   STAR-0: emergence of a nonlanguage content in agent A as the requestor
   STAR-1: production of a request by agent A as the speaker
   STAR-2: interpretation of the request by agent B as the hearer
   STAR-1: nonlang. or lang. fulfillment action by agent B as the requestee
   STAR-2: nonlanguage or language fulfillment recognition by agent A as the requestor

Our analysis has proceeded systematically from the perspective of the speaker to that of the hearer. The transition from one perspective to the next is described in terms of explicit STAR inferences. Part of the analysis is the interpretation of first and second person pronouns (indexicals), which is different for the speaker and the hearer. Our theory complements Schegloff's (2007) sociolinguistic analysis of many recorded and transcribed dialogues.

11. Computing Perspective in Text

Natural language surfaces transport three basic kinds of information. These are (i) the propositional content, or content for short, (ii) the evaluation of the content, also called appraisal, and (iii) the perspectives of the speaker and the hearer onto the content. The transfer of (i), propositional content, has been discussed throughout the previous chapters. The mechanism of appraisal, (ii), for the purpose of maintaining balance has been focused on in Chap. 5. In this chapter we turn to (iii), the computation of perspective in text, complementing the computation of perspective in dialogue analyzed in the previous chapter.

11.1 Coding the STAR-1 in Written Text

Pre-theoretically, there are many different kinds of texts, ranging from rhymed and unrhymed plays to novels, to newspaper articles in such different domains as politics, economics, feuilleton, sports, etc., to journal papers on physics, biology, medicine, mathematics, philosophy, law, religion, etc., to private letters, to the mundane phone books and bank statements. As written signs, they have in common that the STAR-1 parameter values of their speak (write) mode production may be arbitrarily far removed from the STAR-2 parameter values of their hear (read) mode interpretation.

Given that language signs code the speaker's STAR-1 perspective from the utterance situation towards the content's STAR-0 origin (Sects. 10.1, 10.2), it is essential for the hearer to know the utterance situation as specified by the STAR-1. Because the reader (hearer) is normally not present when a text is written – in contradistinction to the hearer in face-to-face communication – it is in the interest of the writer to code the STAR-1 into the text itself. Otherwise, the reader will have difficulty anchoring the written sign to its context of use, which would compromise a correct and complete interpretation.

As an excellent text sample, consider the following citation:

11.1.1 TEXT WITH A DISPERSED CODING OF THE STAR-1

Jan. 16th, 1832 – The neighbourhood of Porto Praya, viewed from the sea, wears a desolate aspect. The volcanic fire of past ages, and the scorching heat of the tropical sun, have in most places rendered the soil steril and unfit for vegetation. The country rises in successive steps of table land, interspersed with some truncate conical hills, and the horizon is bounded by an irregular chain of more lofty mountains. The scene, as beheld through the hazy atmosphere of this climate, is one of great interest; if, indeed, a person, fresh from the sea, and who has just walked, for the first time, in a grove of cocoa-nut trees, can be a judge of anything but his own happiness.

Charles Darwin 1839, *Voyage of the Beagle*, p. 41

How is it possible for the speaker/writer to code a three-dimensional landscape into a one-dimensional sequence of word form surfaces which the hearer/reader can decode into an image of that same three-dimensional landscape in his or her mind? And not only is a landscape described by this string of word forms, but also are the author's feelings of going on land after a long voyage in the confined quarters of the HMS Beagle.[1]

The DBS answer has three parts. The first part is the implementation of concepts for the agent's basic recognition and action procedures, which are reused as the core values in the language proplets constituting the text's content (lexical semantics). The second part is the functor-argument and coordination structure of the text's content, based on the lexical proplet structure of the word forms and their time-linear concatenation (compositional semantics). The third part is the anchoring of the text's content to the STAR-1 of its utterance[2] (pragmatics).

The latter raises the question of where the reader can find the values of a written sign's STAR-1. For a human reader, finding these values in a book is usually easy, even if they are dispersed in various places, as in the above example: The S value Porto Praya of the appropriate STAR-1 is embedded into the first sentence of the text. The T value 1832-01-16 is given as the diary date Jan. 16th, 1832 preceding the text. The A value equals the author, as stated on the book's cover. And the R value is the general readership.

Other possible choices are the S value London and the T value 1839 for where and when, respectively, the book was first published. These may be appropriate for another text, but not for our example. Such ambiguities between appropriate and inappropriate STAR-1 values present a difficulty for an artificial agent. They may be resolved by standardizing the specification of the STAR-1 for newly written texts and by defining templates for different kinds of books and different authors for existing texts – plus human help when needed.

---

[1] The Beagle was a British Cherokee class sailing ship designed in 1807.

[2] I.e., the speaker/writer's perspective from the utterance situation towards the STAR-0 origin of the content.

Often the author's circumstances of writing, coded in the STAR-1, differ from the STAR-0 (Sect. 10.2); the STAR-0 codes the circumstances of a content's origin as a recording of the agent's current recognition and action (Sect. 10.1). However, if the agent writes down the recognition and action data immediately, the STAR-1 and the STAR-0 are practically the same. In this case, the perspective of the STAR-1, relating the utterance situation to the STAR-0 of the content, is shortened to the perspective of the STAR-0.

Mapping a STAR-0 = STAR-1 content into language creates an effect of immediacy and authenticity. This is what the author achieves in example 11.1.1: there is no direct reference to any earlier events or locations of a separate STAR-0 content. Instead, the use of the present tense leads the reader to view Porto Praya with the eyes of the author from the deck of the **Beagle** on Jan. 16th, 1832. The reader can relax and enjoy the author's report because he or she is neither expected nor even able to derive a response which could reach the author, in contradistinction to a face-to-face question or request.

In DBS, a text like 11.1.1 is analyzed as a sequence of statement dialogues. As shown in 10.6.7, each elementary statement dialogue includes the derivation of a STAR-2 perspective by the hearer/reader (otherwise it would be a monologue). In 11.1.1, the STAR-2 perspective is not used as the basis for a subsequent hearer/reader response (in contradistinction to Sect. 10.3). It is not even used for interpreting the indexicals moi and toi (because there are none in the text). It only provides a perspective as an integral part of the reader's understanding, looking back from her or his current STAR-2 circumstances to the author's STAR-1 = STAR-0 circumstances in the year 1832.

While the computation of the speaker/writer and the hearer/reader perspectives based on the STAR-0, STAR-1, and STAR-2 is completely mechanical, there is another aspect of understanding which relates to differences in the background knowledge of different readers (cf. FoCL'99, Sect. 21.5). For example, a reader who knows the location of Porto Praya on the globe will have a better understanding of the sample text than one who does not. Similarly, a reader who has already experienced a walk under cocoa-nut trees, or traveled on a sailing ship will have a more complete understanding of the text than one who has not.

Such differences due to different background knowledge apply to the understanding not only of statement dialogues, but also of question and request dialogues. This kind of unsystematic variation may be easily modeled by providing the Word Banks of different artificial agents with different contents. It is neither an intrinsic part of nor an obstacle to the language communication mechanism reconstructed in DBS.

11.2 Direct Speech in Statement Dialogue

One aspect in which statement dialogues differ from question and request dialogues is the interpretation of **moi** and **toi** in direct (quoted) speech. Consider the following example:

11.2.1 DIRECT SPEECH IN A STATEMENT CONTENT

John said to Mary: *moi* love *toi*. Mary said to John: *moi* love *toi*.[STAR−1]

The indexicals *moi* and *toi* are not interpreted relative to the A and R values of the STAR, in contradistinction to the statement dialogue in Sect. 10.3. Instead, they are interpreted relative to the subject and the object of the sentence into which they are embedded. Consider the following representation of the content 11.2.1 as a set of proplets:

11.2.2 REPRESENTING THE CONTENT OF 11.2.1 AS A SET OF PROPLETS

$$
\begin{bmatrix} \text{noun: John} \\ \text{fnc: say} \\ \text{prn: 23} \end{bmatrix}
\begin{bmatrix} \text{verb: say} \\ \text{arg: John to\_Mary (love 24)} \\ \text{nc: (say 25)} \\ \text{prn: 23} \end{bmatrix}
\begin{bmatrix} \text{adj: to\_Mary} \\ \text{fnc: say} \\ \text{prn: 23} \end{bmatrix}
\begin{bmatrix} \text{noun: moi} \\ \text{fnc: love} \\ \text{prn: 24} \end{bmatrix}
\begin{bmatrix} \text{verb: love} \\ \text{arg: moi toi} \\ \text{fnc: (say 23)} \\ \text{prn: 24} \end{bmatrix}
\begin{bmatrix} \text{noun: toi} \\ \text{fnc: love} \\ \text{prn: 24} \end{bmatrix}
$$

$$
\begin{bmatrix} \text{noun: Mary} \\ \text{fnc: say} \\ \text{prn: 25} \end{bmatrix}
\begin{bmatrix} \text{verb: say} \\ \text{arg: Mary to\_John (love 26)} \\ \text{pc: (say 23)} \\ \text{prn: 25} \end{bmatrix}
\begin{bmatrix} \text{adj: to\_John} \\ \text{fnc: say} \\ \text{prn: 25} \end{bmatrix}
\begin{bmatrix} \text{noun: moi} \\ \text{fnc: love} \\ \text{prn: 26} \end{bmatrix}
\begin{bmatrix} \text{verb: love} \\ \text{arg: moi toi} \\ \text{fnc: (say 25)} \\ \text{prn: 26} \end{bmatrix}
\begin{bmatrix} \text{noun: toi} \\ \text{fnc: love} \\ \text{prn: 26} \end{bmatrix}
$$

The propositions 23 and 25 are connected by the **nc/pc** values of the two *say* proplets. The two matrix propositions treat their quotation like a sentential object (7.5.1, 11.4.5, 11.4.6); the relation to the matrix verb is coded by the additional **fnc** attributes in the two instances of *love*, which take the core and the **prn** value of the associated matrix verb as their value.

For storage in the hearer/reader's Word Bank, the content as specified in 11.2.2 remains unchanged: in contradistinction to 10.3.4, there is no reinterpretation of the *moi* and *toi* indexicals. The STAR-1 attributed by the hearer/reader to the speaker/writer, however, is changed to the STAR-2 of the hearer/reader in accordance with the STAR-2.2 inference defined in 10.3.6.

Nevertheless, in order to properly grasp the content of 11.2.2, the hearer must be able to know who the different occurrences of **moi** and **toi** refer to. In an artificial agent, this may be accomplished by the following inference:

11.2.3 STAR-2.5 INFERENCE INTERPRETING *moi/toi* IN QUOTED SPEECH

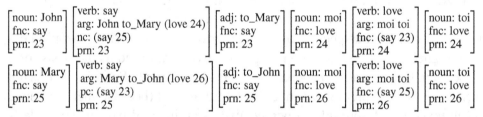

$$
\begin{bmatrix} \text{verb: } \gamma \\ \text{arg: U } \delta \text{ Z } (\beta \text{ K}) \\ \text{prn: K-1} \end{bmatrix}
\begin{bmatrix} \text{noun: } \alpha \\ \text{fnc: } \beta \\ \text{prn: K} \end{bmatrix}
\begin{bmatrix} \text{verb: } \beta \\ \text{arg: X } \alpha \text{ Y} \\ \text{fnc: } (\gamma \text{ K-1}) \\ \text{prn: K} \end{bmatrix}
\quad
\begin{matrix} \text{where } \alpha \in \{\text{moi, toi}\} \\ \text{and } \gamma \in \{\text{say, tell, ...}\} \end{matrix}
\Rightarrow
\begin{matrix} \text{if U = NIL, then } \delta = moi \\ \text{otherwise, } \delta = toi \end{matrix}
$$

When applying this inference to the first complex content in 11.2.2, α may be bound to moi or toi, β is bound to love, and γ is bound to say. These values satisfy the variable restrictions of the inference. Furthermore, δ is bound to *John* if U is bound to NIL or to *Mary* if U is not bound to NIL. As the result, *moi* is identified with John and *toi* with Mary in the first complex content and conversely in the second, in concord with intuition.[3]

In contradistinction to the STAR-2.1 inference defined in 10.3.3, which permanently modifies the input content before storage in the hearer's Word Bank, the STAR-2.5 inference applies after storage without changing the content: The inference is applied whenever a content like 11.2.2 is activated.

The kinds of inferences defined so far may be summarized as follows:

11.2.4 NONLANGUAGE INFERENCES FOR MAINTAINING BALANCE

1. R(eactor) inferences for recognizing an imbalance and deriving a countermeasure (5.2.1–5.2.3).
2. D(eductor) inferences for establishing meaning and event relations (5.3.1–5.3.3).
3. D inferences for creating summaries (5.3.5, 6.5.4, 6.5.5).
4. Consequence inferences CIN and CIP (6.3.6, 6.3.7).
5. Meta-inference for deriving **up** and **down** inferences for hierarchies (6.5.7).
6. The resulting inferences for performing upward and downward traversal in a hierarchy (6.5.9, 6.5.12).
7. E(ffector) inferences for deriving blueprints for action (5.2.1, 5.2.5, 5.5.5).
8. E inference for changing subjunctive to imperative content (5.6.1).

nonlanguage inferences attach to old content (input) by means of pattern matching and derive new content (output), stored at the *now front*, without affecting the old content. Their task is to maintain the agent's equilibrium by recognizing deviations from balance and deriving blueprints for action.

Language inferences, in contrast, adjust the different perspectives of the speaker and the hearer on a content:

11.2.5 LANGUAGE INFERENCES FOR ADJUSTING PERSPECTIVE

1. STAR-1 inferences deriving the speaker's perspective on Spatial, Temporal (10.2.2), and Recipient (10.2.4) aspects of STAR-0 contents.
2. STAR-2 inferences deriving the hearer's perspective on contents with (10.3.4, 10.4.5) and without (10.3.6) 1st and 2nd person indexicals.

---

[3] This inference works even for Then I thought: You don't like this. with You referring to the speaker.

3. STAR-2 inference combining question with answer content (10.4.10).
4. STAR-2 inference interpreting *moi/toi* in quoted speech (11.2.3).
5. STAR-2 inference interpreting ça coreferentially (11.3.6)

The STAR-1 inferences derive a modified version of a STAR-0 content which is stored at the *now front* of the speaker's Word Bank and used for language production. The STAR-2.1–STAR-2.4 inferences change a STAR-1 content prior to storage at the *now front* of the hearer's Word Bank. The STAR-2.5 inference 11.2.3 does not derive new content, but applies to stored content without permanently changing it – as if it were looking through a pair of glasses.

The inferences listed in 11.2.4 and 11.2.5 all use the same format, consisting of an input schema, a connective, an output schema, and variable restrictions. It is a matter of further research to design more inferences as needed, to combine them to obtain generalizations where possible, and to classify them into more and more differentiated groups and subgroups. According to the Sequential Inferencing Principle (SIP, 5.2.5), two inferences x and y may be applied in sequence iff the consequent of x equals the antecedent of y.

11.3 Indexical vs. Coreferential Uses of 3rd Pronouns

In texts containing direct speech, moi and toi refer to language proplets – instead of pointing to the values of the STAR's A and R attributes, as in dialogue. Such reference to activated language items, based on the inference 11.2.3, provides us with a model for the interpretation of third person indexicals.

The pointer ça may refer not only indexically to the value of the STAR's 3rd attribute (e.g., 10.1.6), but also coreferentially to activated language items, assuming grammatical and semantic compatibility. Consider the following example of an object sentence construction:

11.3.1 Ambiguous hear mode content

Mary knew that *she* was happy.^{STAR-2}

On the indexical reading, *she* refers to a female other than Mary, e.g., Suzy; this may be shown graphically as follows:

11.3.2 Indexical use of she

Mary   knew that   she   was happy.

On the coreferential reading, *she* refers to Mary (repeated reference, cf. FoCL'99, Sect. 6.3); this may be shown graphically as follows:

11.3.3 COREFERENTIAL USE OF she

In linguistics, the "full" noun referred to by a coreferential pronoun is called the *antecedent* if it precedes, and the *postcedent* if it follows the pronoun.

The ambiguity between the indexical and the coreferential readings arises in STAR-2 contents after parsing the incoming surface and before adjusting to the hearer perspective (Sect. 10.3). No such ambiguity arises in a corresponding STAR-0 content recording the agent's current observation because it is the speaker who selects the intended reference (compare 10.1.5 and 10.1.6).

An indexical reference[4] to Suzy is illustrated by the interpretation of 11.3.1 as a STAR-0 content, showing the co-indexed STAR-0 proplet at the end:

11.3.4 INDEXICAL STAR-0 REPRESENTATION AS A PROPLET SET

noun: Mary cat: nm sem: sg f fnc: know prn: 89	verb: know cat: decl sem: past arg: Mary (happy 90) prn: 89	noun: ça cat: s3 sem: sg f fnc: happy prn: 90	verb: happy cat: v sem: past arg: ça fnc: (know 89) prn: 90	S: Austin T: 1974-09-12 A: Peter 3rd: Suzy prn: 89

The STAR-0 proplet anchors the proposition to the circumstances in which author Peter records the content. The crucial content value is the indexical ça, which points to the 3rd attribute with the value Suzy of the STAR-0 proplet. When the agent computes a STAR-1 perspective towards this content for language production, the indexical interpretation is realized as the pronoun she.

Next consider the coreferential interpretation of 11.3.1 as a STAR-0 content:

11.3.5 COREFERENTIAL STAR-0 REPRESENTATION AS A PROPLET SET

noun: Mary cat: nm sem: sg f fnc: know prn: 93	verb: know cat: decl sem: past arg: Mary (happy 94) prn: 93	noun: ça cat: s3 sem: sg f fnc: happy prn: 94	verb: happy cat: v sem: past arg: ça fnc: (know 93) prn: 94	S: Austin T: 1974-09-12 A: Peter prn: 93

Here the STAR-0 contains no 3rd value for the interpretation of ça. Therefore, this indexical is interpreted by the following inference:

---

[4] Bosch (1986) uses the notions "syntactic reference" for our coreferential reference and "referential reference" for our indexical reference.

11.3.6 INFERENCE FOR THE COREFERENTIAL INTERPRETATION OF ça

$$\begin{bmatrix} \text{noun: } \alpha \\ \text{prn: K} \end{bmatrix} \cdots \begin{bmatrix} \text{noun: ça} \\ \text{prn: K+M} \end{bmatrix} \Longrightarrow \begin{bmatrix} \text{noun: } (\alpha \text{ K}) \\ \text{prn: K+M} \end{bmatrix}$$

Applied to 11.3.5, the inference replaces the value **ça** with the address **(Mary 93)**, establishing coreference with the first proplet. When the agent computes a STAR-1 perspective towards this content for language production, the coreferential interpretation is maintained and **(Mary 93)** is realized as the pronoun **she**.

   Two different contents for the speaker, namely 11.3.4 and 11.3.5, with the same surface create an ambiguity for the hearer. When computing the STAR-2 perspective on the content derived from the ambiguous surface, the hearer must make a choice and try to select the interpretation intended by the speaker. There are three possibilities: the hearer selects (i) the same interpretation as the speaker (correct interpretation), (ii) a different interpretation (incorrect interpretation), or (iii) remains undecided (no complete interpretation). In the latter case there may be the option to request clarification from the speaker.

11.4 Langacker-Ross Constraint for Sentential Arguments

It has been observed by Langacker (1969) and Ross (1969) that 3rd person pronouns in certain subclause constellations can *only* have an indexical interpretation. This applies to all kinds of subclause constructions, i.e., (i) the subject sentence, (ii) the object sentence, (iii) the adnominal modifier sentence (aka relative clause), and (iv) the adverbial modifier sentence.

   Consider the following variants of a subject sentence construction:

11.4.1 PRONOUN IN SUBJECT SENTENCE CONSTRUCTIONS

1. LH' Coreferent noun in lower clause (L) precedes pronoun in matrix (H')
        That *Mary* was happy surprised *her*.
2. H'L Pronoun in matrix (H') precedes non-coref. noun in lower clause (L)
        % *She* was surprised that *Mary* was happy.
3. L'H Pronoun in lower clause (L') precedes coreferent noun in matrix (H)
        That *she* was happy surprised *Mary*.
4. HL' Coreferent noun in matrix (H) precedes pronoun in lower clause (L')
        Mary was surprised that *she* was happy.

All four sentences are wellformed. They differ as to (i) whether the lower clause L (subclause) precedes (LH) or follows (HL) the higher clause H (matrix), and (ii) whether the pronoun is in the lower clause (L') or in the higher clause (H'). The constructions 1, 3, and 4 are alike in that they permit an in-dexical as well as a coreferential interpretation. The construction 2, in contrast, allows only an indexical interpretation (as shown graphically in 11.3.2); this more restricted interpretation of construction 2 is indicated by the % marker.

Given that the choice of a pronoun and its form of reference (indexical vs. coreferential) originate in the speaker, let us consider the DBS graph analyses of these constructions. We begin with the first two examples of 11.4.1, which have the pronoun in the higher clause. They share the SRG, but have two different NAGs and two different surface realizations, (a) and (b):

11.4.2 SUBJECT SENTENCE: PRONOUN IN HIGHER CLAUSE (MATRIX)

1. LH': That *Mary* was happy surprised *her*.
2. H'L: % *She* was surprised that *Mary* was happy.

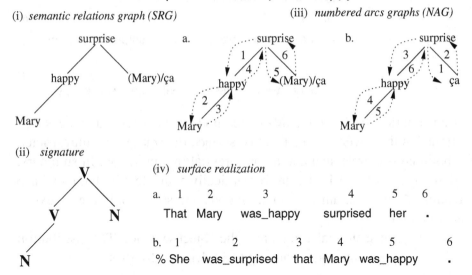

(i) *semantic relations graph (SRG)*   (iii) *numbered arcs graphs (NAG)*

(ii) *signature*

(iv) *surface realization*

For ease of exposition, the node allowing a coreferential as well as an indexical interpretation is represented as **(Mary)/ça. (Mary)** represents an address and stands for the speaker's choice of a coreferential reference (as in 11.3.5), while **ça** stands for the choice of an indexical reference (as in 11.3.4).

NAG (a) represents the object of the higher clause as **(Mary)/ça**, like the SRG. Consequently, the surface realization may have (i) an indexical or (ii) a coreferential pronoun interpretation for the hearer. The navigation represented by NAG (b), in contrast, does not allow a coreferential pronoun interpretation. Therefore the node is represented as unambiguous **ça** and the surface realiza-tion (b) has only an indexical pronoun interpretation, indicated by %.

Next consider the DBS graph analysis of examples 3 and 4 in 11.4.1. As in 11.4.2, they have two different NAGs and two different surface realizations.

11.4.3 SUBJECT SENTENCE: PRONOUN IN LOWER CLAUSE

3. L'H: That *she* was happy surprised *Mary*.
4. HL': *Mary* was surprised that *she* was happy.

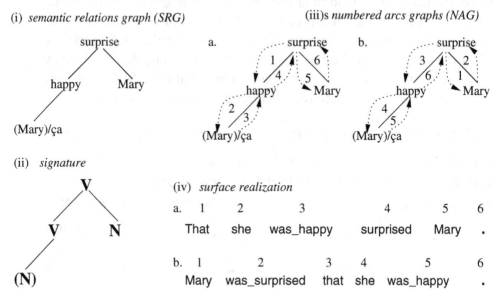

(i) *semantic relations graph (SRG)*

(ii) *signature*

(iii)s *numbered arcs graphs (NAG)*

(iv) *surface realization*

a.	1	2	3	4	5	6
	That	she	was_happy	surprised	Mary	.

b.	1	2	3	4	5	6
	Mary	was_surprised	that	she	was_happy	.

In contradistinction to 11.4.2, the **(Mary)/ça** node in the SRG and in the NAGs is positioned in the lower clause. The two surface realizations are unrestricted in that both have an indexical and a coreferential interpretation. In summary, the variants 1, 3, and 4 in 11.4.1 are semantically equivalent in that they have both readings, while variant 2 is unambiguous and shares only the indexical reading with them.

The other kind of sentential argument is the object sentence. The distribution of coreferential pronouns corresponding to 11.4.1 is as follows:

11.4.4 PRONOUN IN OBJECT SENTENCE CONSTRUCTIONS

1. LH'   Coreferent noun in lower clause (L) precedes pronoun in matrix (H')
         That *Mary* was happy was known to *her*.
2. H'L   Pronoun in matrix (H') precedes non-coref. noun in lower clause (L)
         % *She* knew that *Mary* was happy.
3. L'H   Pronoun in lower clause (L') precedes coreferent noun in matrix (H)
         That *she* was happy was known to *Mary*.
4. HL'   Coreferent noun in matrix (H) precedes pronoun in lower clause (L')
         Mary knew that *she* was happy.

As in 11.4.1, the alternative ordering of higher and lower clause is based on using passive (9.1.7) instead of active (9.1.3) in the higher clause. Languages with a freer word order than English may use fronting of the object sentence as an alternative to a passive construction.[5]

Consider the DBS graph analysis of examples 1 and 2:

11.4.5 OBJECT SENTENCE: PRONOUN IN HIGHER CLAUSE (MATRIX)

1. LH': That *Mary* was happy was known to *her*.
2. H'L: % *She* knew that *Mary* was happy.

(i) *semantic relations graph (SRG)*   (iii) *numbered arcs graph (NAG)*

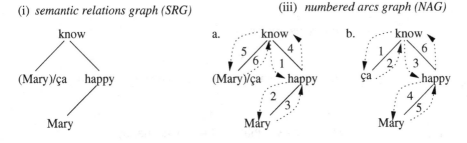

(ii) *signature*

(iv) *surface realization*

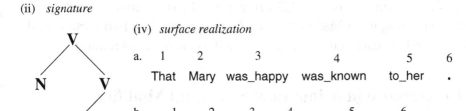

In the SRG and NAG (a), the subject of the higher clause is represented by the ambiguous **(Mary)/ça** node. In NAG (b), in contrast, this node is represented unambiguously as **ça** because this navigation permits only an indexical pronoun interpretation. This is similar to NAG (b) in the subject sentence constructions 11.4.2.

Next consider the DBS graph analysis of constructions 3 and 4 listed in 11.4.4 with the pronoun in the lower clause:

11.4.6 OBJECT SENTENCE: PRONOUN IN LOWER CLAUSE

3. L'H: That *she* was happy was known to *Mary*.
4. HL': *Mary* knew that *she* was happy.

---

[5] For example, sentence 1 in 11.4.4 may be translated into German as Dass sich Maria freute, wurde von ihr bemerkt. (passive) or Dass sich Maria freute, bemerkte sie. (fronting of object sentence).

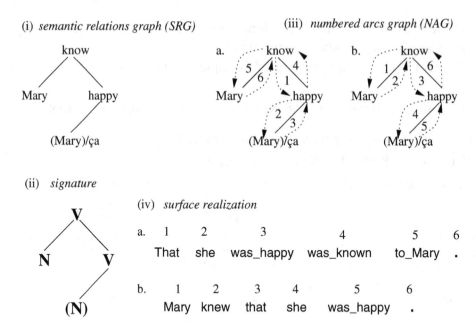

(i) *semantic relations graph (SRG)*

(iii) *numbered arcs graph (NAG)*

(ii) *signature*

(iv) *surface realization*

a.

1	2	3	4	5	6
That	she	was_happy	was_known	to_Mary	.

b.

1	2	3	4	5	6
Mary	knew	that	she	was_happy	.

As in 11.4.3, the traversals of NAG (a) and NAG (b) are unrestricted in that both have an ambiguous **Mary/ça** node. Consequently, the two surface realizations (a) and (b) each have an indexical and a coreferential reading.

11.5 Coreference in Adnominal Sentential Modifiers

In English, sentential adnominal modifiers (aka relative clauses) are positioned directly after their modified ("head noun") – unless they are extraposed (Sect. 9.3). Therefore, the subclause may be located in the middle of the main clause (cf. 1 and 3 in 11.5.1), unlike sentential subjects and objects. However, because the restriction observed by Langacker and Ross applies to the relative order of the pronoun and the coreferent noun (rather than the order of the subclause and the main clause), the constraint applies also to relative clauses:

11.5.1 PRONOUN IN ADNOMINAL MODIFIER CONSTRUCTIONS

1. LH' Coreferent noun in lower clause (L) precedes pronoun in matrix (H')
    The man who loves *the woman* **kissed** *her.*
2. H'L Pronoun in matrix (H') precedes non-coref. noun in lower clause (L)
    % *She* **was kissed by the man who loves** *the woman.*
3. L'H Pronoun in lower clause (L') precedes coreferent noun in matrix (H)
    The man who loves *her* **kissed** *the woman.*
4. HL' Coreferent noun in matrix (H) precedes pronoun in lower clause (L')
    The woman **was kissed by the man who loves** *her.*

The four constructions are analogous to those of 11.4.1 and 11.4.4. Again, it is the H'L constellation which is restricted to indexical reference.

The first two examples have the following DBS graph analysis:

11.5.2 ADNOMINAL MODIFIER SENTENCE: PRONOUN IN HIGHER CLAUSE

1. LH': The man who loves *the woman* kissed *her*.
2. H'L: % *She* was kissed by the man who loves *the woman*.

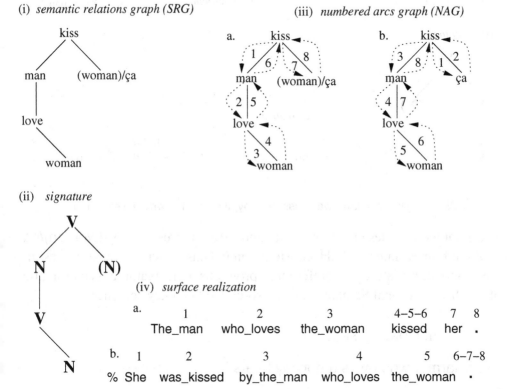

(i) *semantic relations graph (SRG)*

(iii) *numbered arcs graph (NAG)*

(ii) *signature*

(iv) *surface realization*

a.
1	2	3	4–5–6	7	8
The_man	who_loves	the_woman	kissed	her	.

b.
1	2	3	4	5	6–7–8
% She	was_kissed	by_the_man	who_loves	the_woman	·

As in the analyses of sentential arguments (Sect. 11.4), the ambiguous node in NAG (a) is represented as (woman)/ça, while the unambiguous node in NAG (b) is represented as ça.

The following constructions with the pronoun in the lower clause, in contrast, have ambiguous nodes in both NAGs:

11.5.3 ADNOMINAL MODIFIER SENTENCE: PRONOUN IN LOWER CLAUSE

3. L'H: The man who loves *her* kissed *the woman*.
4. HL': *The woman* was kissed by the man who loves *her*.

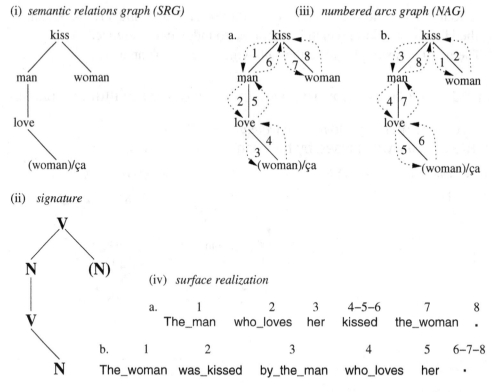

(i) *semantic relations graph (SRG)*

(iii) *numbered arcs graph (NAG)*

(ii) *signature*

(iv) *surface realization*

	a.	1	2	3	4–5–6	7	8
		The_man	who_loves	her	kissed	the_woman	.

	b.	1	2	3	4	5	6–7–8
		The_woman	was_kissed	by_the_man	who_loves	her	.

The ambiguous nodes in NAGs (iii) (a) and (b) are represented by (woman)/ça.

Of the four variants, the LH' constellation (1) has drawn the most interest in the abundant literature on coreferential pronouns. One example, symptomatic of Truth-Conditional Semantics, is the so-called "Donkey sentence:"

11.5.4 THE DONKEY SENTENCE

Every farmer who owns a donkey beats it.

The seemingly most natural translation into a formula of Truth-Conditional Semantics (i.e., predicate calculus used for the analysis of natural language meaning) is the following:

11.5.5 QUANTIFIER STRUCTURE ATTRIBUTED TO A DONKEY SENTENCE

$\forall x\ [[\text{farmer}(x) \land \exists y\ [\text{donkey}(y) \land \text{own}(x,y)] \rightarrow \text{beat}(x,y)]$

Unfortunately, the y in beat(x,y) is not in the scope of the quantifier $\exists y$ binding donkey(y) in the subordinate clause, as pointed out by Geach (1969).[6]

---

[6] Discourse Representation Theory (DRT) by Kamp (cf. Kamp 1981, Kamp and Reyle 1993, Geurts 2002) originated as an attempt to resolve this well-know problem of Truth-Conditional Semantics.

This, however, is not even the main problem of Truth-Conditional Semantics for building a robot. The fatal inadequacy is a methodological one: definitions in a metalanguage cannot be understood by a computer. What is needed instead are procedural definitions, suitable for implementation.

For example, **Every farmer snores** would be analyzed as

1. $\forall x[\text{farmer}(x) \rightarrow \text{snore}(x)]$.

This expression is formally interpreted relative to a set-theoretically defined model @ and a variable assignment **g**, superscripted at the end as in

2. $\forall x[\text{farmer}(x) \rightarrow \text{snore}(x)]^{@,g}$.

According to the metalanguage definition of the universal quantifier, formula (2) is true with respect to the model @ if the formula without the universal quantifier,[7] i.e.,

3. $[\text{farmer}(x) \rightarrow \text{snore}(x)]^{@,g'}$,

is true with respect to the model @ and *all* assignments **g'** to x – usually infinitely many. This method rests entirely on understanding the meaning of the word *all* in the metalanguage.[8]

The DBS alternative is metalanguage-independent patterns which characterize **all** as exhaustive, **some** as selective, and so on (cf. NLC'06, 6.2.9). Using the pattern names as proplet values, the "quantifiers" are redefined as determiners. Instead of restructuring the content of 11.5.4 beyond recognition (and without success), as in 11.5.5, the DBS alternative is an automatic, time-linear, surface compositional hear mode derivation. The quantifiers binding variables horizontally, causing the scope problem in 11.5.5, are set aside and replaced by addresses which establish coreference between **it** and **donkey** successfully:

11.5.6 DBS GRAPH ANALYSIS OF THE DONKEY SENTENCE

(i) *semantic relations graph (SRG)*   (iii) *numbered arcs graph (NAG)*

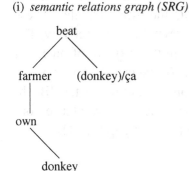
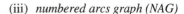
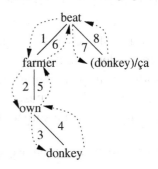

---

[7] The purpose is to reduce the interpretation to the truth conditions of propositional calculus.

[8] Other reasons why Truth-Conditional Semantics is unsuitable for building robots are the [-sense, -constructive] ontology on which it is based (cf. FoCL'99, Sect. 20.4), and the resulting absence of recognition and action procedures as well as a memory in an agent.

(ii) *signature*

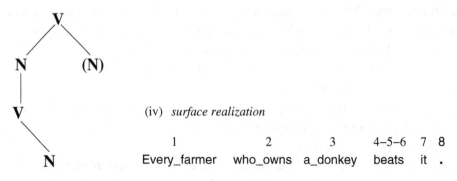

(iv) *surface realization*

	1	2	3	4-5-6	7	8
	Every_farmer	who_owns	a_donkey	beats	it	.

The redefinition of quantifiers as **sem** values and the coreference between **donkey** and **it** are shown by the following set of proplets, which codes the content of 11.5.4 in DBS:

11.5.7 REPRESENTING THE DONKEY CONTENT AS A SET OF PROPLETS

$$
\begin{bmatrix} \text{noun: farmer} \\ \text{cat: snp} \\ \text{sem: pl exh} \\ \text{fnc: beat} \\ \text{mdr: (own 17)} \\ \text{prn: 16} \end{bmatrix}
\begin{bmatrix} \text{verb: own} \\ \text{cat: v} \\ \text{sem: pres} \\ \text{arg: \# donkey} \\ \text{mdd: (beat 16)} \\ \text{prn: 17} \end{bmatrix}
\begin{bmatrix} \text{noun: donkey} \\ \text{cat: snp} \\ \text{sem: indef sg} \\ \text{fnc: own} \\ \text{prn: 17} \end{bmatrix}
\begin{bmatrix} \text{verb: beat} \\ \text{cat: decl} \\ \text{sem: pres} \\ \text{arg: farmer (donkey 17)} \\ \text{prn: 16} \end{bmatrix}
\begin{bmatrix} \text{noun: (donkey 17)} \\ \text{cat: snp} \\ \text{sem: indef sg} \\ \text{fnc: beat} \\ \text{prn: 16} \end{bmatrix}
$$

The noun proplet *farmer* has the **sem** values **pl exh** (plural exhaustive) in combination with the **cat** value **snp** (singular noun phrase). The noun proplet *donkey* has the cat value **snp** in combination with the **sem** values **indef sg** (indefinite singular). The coreference of **it** with the antecedent **donkey** is coded by the rightmost proplet with the address core value **(donkey 17)**.

The main clause and the relative clause have different **prn** values, here **16** and **17**. The modifier|modified relation between the subclause verb *own* and the main clause noun *farmer* is coded by their **mdd** and **mdr** attributes. These attributes have the address values **(beat 16)** and **(own 17)**, respectively. The subject gap of the relative clause is indicated by # in the **arg** slot of *own*. For hear mode derivations of relative clauses, see NLC'06, Sects. 7.4 and 7.5.

Another example, symptomatic of Transformational Grammar, is the "Bach-Peters sentence." It is based on intertwined L'H and HL' relative clause constructions, one with a subject gap, the other with an object gap (9.3.1):

11.5.8 THE BACH-PETERS SENTENCE

The man who deserves it will get the prize he wants.[9]

---

[9] Grammatically equivalent is The pilot who shot at it hit the MIG that chased him. (Karttunen 1971).

The difficulty arises if the pronoun it is derived transformationally from the "underlying" noun phrase the prize he wants, and the pronoun he from the man who deserves it. Because each underlying noun phrase contains a pronoun which itself is based on an underlying noun phrase there result two recursions leading to an infinite "deep structure." Based on this Phrase Structure Grammar analysis, Peters and Ritchie (1973) proved that transformational grammar is undecidable (complexity).

The alternative DBS analysis of this example is surprisingly simple, like that of the center-embedded relative clauses shown in 9.3.3:

11.5.9 DBS GRAPH ANALYSIS OF THE BACH-PETERS SENTENCE

(i) *semantic relations graph (SRG)*   (iii) *numbered arcs graph (NAG)*

   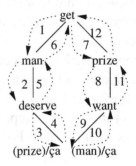

(ii) *signature*

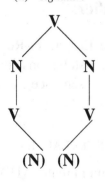

(iv) *surface realization*

1		2	3	4–5–6	7	8–9	10	11–12
The_man	who_deserves		it	will_get	the_prize	he	wants	.

The graph indicates that the first relative clause who deserves it has a subject gap, while the second relative clause (which) he wants has an object gap. For the hear mode derivation of this example, see NLC'06, 10.6.2.

The representation of the content as a set of proplets is as follows:

11.5.10 PROPLET REPRESENTATION OF THE BACH-PETERS SENTENCE

$$
\begin{bmatrix} \text{noun: man} \\ \text{cat: snp} \\ \text{sem: def sg} \\ \text{fnc: get} \\ \text{mdr: (deserve 57)} \\ \text{prn: 56} \end{bmatrix}
\begin{bmatrix} \text{verb: deserve} \\ \text{cat: v} \\ \text{sem: ind pres} \\ \text{arg: \# (prize 56)} \\ \text{mdd: (man 56)} \\ \text{prn: 57} \end{bmatrix}
\begin{bmatrix} \text{noun: (prize 56)} \\ \text{cat: snp} \\ \text{sem: def sg} \\ \text{fnc: deserve} \\ \text{prn: 57} \end{bmatrix}
\begin{bmatrix} \text{verb: get} \\ \text{cat: decl} \\ \text{sem: pres} \\ \text{arg: man prize} \\ \text{prn: 56} \end{bmatrix}
$$

$$
\begin{bmatrix} \text{noun: prize} \\ \text{cat: snp} \\ \text{sem: def sg} \\ \text{fnc: get} \\ \text{mdr: (want 58)} \\ \text{prn: 56} \end{bmatrix}
\begin{bmatrix} \text{noun: (man 56)} \\ \text{cat: snp} \\ \text{sem: def sg} \\ \text{fnc: want} \\ \text{prn: 58} \end{bmatrix}
\begin{bmatrix} \text{verb: want} \\ \text{cat: v} \\ \text{sem: pres} \\ \text{arg: (man 56) \#} \\ \text{mdd: (prize 56)} \\ \text{prn: 58} \end{bmatrix}
$$

The semantic relation between the first main clause noun proplet and the first relative clause is coded by the modifier value (**deserve 57**) in the **mdr** attribute of *man* and the modified value (**man 56**) in the **mdd** attribute of *deserve*. The pronoun it is represented by the noun proplet with the address (**prize 56**) as its core value.

The semantic relation between the second main clause noun proplet and the second relative clause is coded by the modifier value (**want 58**) in the **mdr** attribute of *prize* and the modified value (**prize 56**) in the **mdd** attribute of *want*. The pronoun **he** is represented by the noun proplet with the address (**man 56**) as its core value. For a more general account see 11.6.5 following.

11.6 Coreference in Adverbial Sentential Modifiers

The four remaining subclause constructions subject to the Langacker-Ross constraint are adverbial modifier sentences. The following constellations correspond to those of 11.4.1 (subject sentence), 11.4.4 (object sentence), and 11.5.1 (adnominal modifier sentence):

11.6.1 LANGACKER-ROSS CONSTRAINT IN ADVERBIAL SUBCLAUSES

1. LH' Coreferent noun in lower clause (L) precedes pronoun in matrix (H')
   When *Mary* returned *she* kissed John.
2. H'L Pronoun in matrix (H') precedes non-coref. noun in lower clause (L)
   % *She* kissed John when *Mary* returned.
3. L'H Pronoun in lower clause (L') precedes coreferent noun in matrix (H)
   When *she* returned *Mary* kissed John.
4. HL' Coreferent noun in matrix (H) precedes pronoun in lower clause (L')
   Mary kissed John when *she* returned.

Due to the relatively free positioning of adverbial modifiers in English, the examples do not require the use of passive for constructing the constellations relevant for the Langacker-Ross constraint – in contradistinction to the constructions analyzed in the two previous sections.

The examples 1 and 2 are based on the following DBS graph analysis:

11.6.2 ADVERBIAL MODIFIER SENTENCE: PRONOUN IN HIGHER CLAUSE

1. LH': When *Mary* returned *she* kissed John.
2. H'L: % *She* kissed John when *Mary* returned.

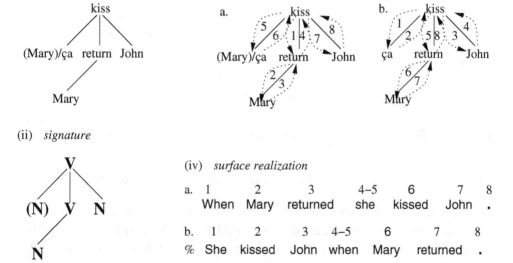

(i) *semantic relations graph (SRG)*

(iii) *numbered arcs graph (NAG)*

(ii) *signature*

(iv) *surface realization*

a.

1	2	3	4–5	6	7	8
When	Mary	returned	she	kissed	John	.

b.

1	2	3	4–5	6	7	8
% She	kissed	John	when	Mary	returned	.

As expected, the **H'L** NAG (b) and the associated surface realization (b) are limited to an indexical interpretation of the pronoun. The content represented by the SRG may be shown as the following set of proplets:

11.6.3 REPRESENTING VARIANT 1 AS A SET OF PROPLETS

$$
\begin{bmatrix} \text{noun: Mary} \\ \text{cat: nm} \\ \text{sem: sg f} \\ \text{fnc: return} \\ \text{prn: 39} \end{bmatrix}
\begin{bmatrix} \text{verb: return} \\ \text{cat: v} \\ \text{sem: past} \\ \text{arg: Mary} \\ \text{mdd: (kiss 40)} \\ \text{prn: 39} \end{bmatrix}
\begin{bmatrix} \text{noun: (Mary 39)/ça} \\ \text{cat: nm} \\ \text{sem: sg f} \\ \text{fnc: kiss} \\ \text{prn: 40} \end{bmatrix}
\begin{bmatrix} \text{verb: kiss} \\ \text{cat: decl} \\ \text{sem: past} \\ \text{arg: (Mary 39)/ça John} \\ \text{mdr: (return 39)} \\ \text{prn: 40} \end{bmatrix}
\begin{bmatrix} \text{noun: John} \\ \text{cat: nm} \\ \text{sem: sg m} \\ \text{fnc: kiss} \\ \text{prn: 40} \end{bmatrix}
$$

The ambiguity of the SRG content, realized by the **LH'** variant 1 and based on the NAG (a) traversal, is coded into the ambiguous value **(Mary 39)/ça**. The set of proplets represents an unanchored content because there is no STAR.

As expected, the examples 3 and 4 allow an indexical as well as a coreferential interpretation:

11.6.4 ADVERBIAL MODIFIER SENTENCE: PRONOUN IN LOWER CLAUSE

3. L'H: When *she* returned *Mary* kissed John.
4. HL': *Mary* kissed John when *she* returned.

(i) *semantic relations graph (SRG)*

(iii) *numbered arcs graph (NAG)*

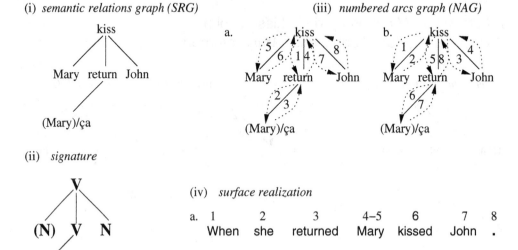

(ii) *signature*

(iv) *surface realization*

a.

1	2	3	4–5	6	7	8
When	she	returned	Mary	kissed	John	.

b.

1	2	3	4–5	6	7	8
Mary	kissed	John	when	she	returned	.

In summary, a coreferential interpretation in subject clause, object clause, adnominal modifier clause, and adverbial modifier clause constructions is possible under the following conditions: (i) the pronoun follows the potentially coreferential noun, as in **LH'** and **HL'**, or (ii) the pronoun is in the lower clause, as in **L'H**. Conversely, a coreferential interpretation is excluded if the pronoun does not follow the potential coreferential noun and is not in the lower clause, as in **H'L**.

The facts are clear, but what is the explanation? Based on the inference 11.3.6, the DBS analysis treats coreferential third person pronouns as address values, e.g., **(Mary)**. The traversal and activation of an address value is successful only if the item to be addressed is available, i.e., has already been activated by a prior traversal step.[10]

To show that a coreferential noun is available to be pointed at by the address value in **LH'** constructions, but not in **H'L** constructions, we must look

---

[10] Due to the strictly time-linear derivation order of LA-grammar, a delayed realization of an address value is not allowed.

at their proplet NAGs (7.4.2). The following examples use an object sentence construction, but the explanation holds for all the other constructions as well:

11.6.5 PROPLET NAGS OF AN H'L AND AN L'H CONSTRUCTION

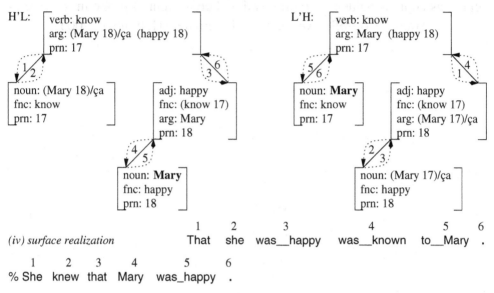

(iv) surface realization

	1	2	3	4	5	6
	That	she	was_happy	was_known	to_Mary	.

	1	2	3	4	5	6
%	She	knew	that	Mary	was_happy	.

For the sake of the argument, the noun proplet representing she in the H'L construction on the left has the core value (Mary 18)/ça rather than only ça – as if a coreferential interpretation were possible. We would like to show why this is in fact not the case.

When the matrix verb *know* is traversed in the initial navigation step, its intrapropositional **arg** values are (sub)activated as well (Sect. 5.4). In the L'H construction on the right of 11.6.5, the relevant **arg** value is intrapropositional Mary. In the H'L construction on the left, in contrast, the relevant **arg** value is an extrapropositional address, namely (Mary 18).

The **arg** value (Mary 18) in the matrix verb of the H'L construction fails as a potential item to be pointed at by the core value (Mary 18) of the noun proplet representing she because it is in need of a value to point at itself. This value would be the *Mary* proplet in the subclause, to be traversed later. Thus, there is no address to point at, and no coreferential interpretation is possible.

In the L'H construction on the right of 11.6.5, in contrast, the postcedent of she, i.e., the *Mary* proplet serving as the higher subject, is activated as soon as the matrix verb is traversed. Thus when the noun proplet representing she is reached in transition 2, there is an item available to be pointed at by the address value (Mary 17) – which explains why a coreferential interpretation is possible in this case.

This concludes the systematic reconstruction of the Langacker-Ross constraint for all four constellations in all four kinds of extrapropositional functor-argument constructions. The constraint is explained in terms of the *availability* of an ante-/postcedent for the coreferential interpretation of pronouns, represented as address values. Intrapropositional coreferential pronoun interpretations, for example, of reflexives, are another matter (Helfenbein 2005).

Part III

Final Chapter

12. Conclusion

For natural language communication to be comfortable, the use of the signs must be *flexible* (robust). Therefore, one challenge in building a talking robot is to reproduce the flexibility which is integral to human communication.

In DBS, this flexibility has three aspects. The first is the ability (i) to change levels of abstraction effortlessly and (ii) to use language indirectly as in analogies, metaphors, and so on. This aspect is based on inferences. Derived and applied automatically, the inferences are triggered by the evaluated data stream provided by the agent's cognition.

The second aspect inheres in the nature of pattern matching. This aspect of flexibility is handled by schemata which utilize restricted variables, core values, and the semantic relations of structure, enabling highly differentiated retrieval. Pattern matching is used, for example, in recognition, when raw data are classified by concept types, and in retrieval, when schemata activate relevant contents in the database.

The third aspect is the ability of the hearer to ask for clarification. This is another aspect handled by inferences.

These aspects of flexibility result combined in practically unlimited expressive power, for example, in data mining. In an alert DBS robot, recognition, subactivation, intersection, and inferencing are continuously running and provide autonomous control with the stored world knowledge and the experiences relevant for maintaining the agent in a state of balance.

12.1 Level of Abstraction

The construction of an artificial agent with language may be compared to the construction of an airplane. For both, there is a natural prototype with which they share certain crucial properties and from which they differ in others.

An airplane (artificial model) and a bird (natural prototype) have in common that both stay airborne according to the same principles of the theory of aerodynamics. They differ, however, in that lift and propulsion are combined in the flapping wings of birds, but separated in planes into the fixed wing for lift and the propeller or jet engine for propulsion.

Correspondingly, an artificial DBS agent and its human prototype have in common that both have a memory connected to external interfaces. An example of their many differences, in contrast, is that DBS uses the equality of core values for sorting proplets into certain token lines of a Word Bank, while the natural counterpart presumably uses a more differentiated similarity metric and different principles of storage and retrieval.

Orthogonal to this analogy between artificial flying and artificial cognition is a crucial difference regarding their interaction with the human user. In an airplane, the method of being airborne is completely separate from its user-friendliness for humans. The latter concerns the size of the doors and the seats, the cabin pressure, the service, etc., while the former concerns the shape of the wings, the manner of propulsion, the technique of takeoff and landing, etc.

For DBS as a computational theory of cognition, in contrast, maximizing the similarity between the artificial agent and its natural prototype amounts directly to maximizing the user-friendliness for humans.[1] This correlation between the artificial agent and its natural prototype has been preformulated as the Equation Principle in NLC'06, 1.3.1: [2]

12.1.1 THE EQUATION PRINCIPLE OF DATABASE SEMANTICS

1. The more realistic the reconstruction of natural cognition, the better the functioning of the artificial model.

2. The better the functioning of the artificial model, the more realistic the reconstruction of natural cognition.

The principle is aimed at long-term upscaling. It applies at a level of abstraction at which it does not matter for completeness of function and data coverage whether cognition is based on natural wetware or electronic hardware.[3] The principle applies to the communication and reasoning aspects[4] of a talking agent which are (i) concretely observable and (ii) impact directly the quality of free human-machine communication in natural language.

---

[1] Without giving up any of the applications provided by computers already.

[2] The principles presented in Chap. 1 of NLC'06 are (1) the Verification Principle, (2) the Equation Principle (12.1.1), (3) the Objectivation Principle, (4) the Equivalence Principle for Interfaces (2.6.2), and (5) the Equivalence Principle for Input/Output (2.6.3).

[3] As for "true feelings," they are equally inaccessible in natural and artificial agents. In fact, they may be more accessible technically in artificial agents due to their service channel (Chap. 2).

[4] It applies less to other aspects of the artificial agent, such as looks – as shown by C3PO and Ripley.

The flexibility required by the human user depends in part on what the partners in discourse accept as communication. For example, high brow journals will accept only papers which exhibit the correct use of the current buzz words in their domain, some bureaucrats are interested only in a small range of topics, such as last name, first name, date of birth, and place of birth, the customers in a bar, who will accept a stranger only if (s)he is a certain type, and so on. In linguistics, these restrictions of a domain to a certain, well-defined protocol are called *register* (Halliday and Hasan 1976). Register is important for communication, natural or artificial, because it defines the channel of communication from a social point of view.

In this sense, the construction of a talking robot must solve the task of *register adaptation.* The system should be able to smoothly agree with the partner in discourse on a certain level of abstraction, to switch the level of abstraction up or down, accompanied by adjustments of dialect and of intonation, to select between a declarative, interrogative, or imperative sentential mood, to present content in a certain way, etc. This is the most fertile field of Conversation Analysis, founded by Sacks and Schegloff (Schegloff 2007).[5]

The DBS robot requires a computational model of the motivational structures behind the actions of the partners in discourse and of the strategies guiding these actions. For this, the Schegloff corpus provides many authentic examples. A DBS interpretation of the Schegloff corpus would have to translate such founding notions as "pre," "post," "pre-pre," etc., (time-linear!) into DBS inferences with certain goals.

To overcome the inflexibility of current systems. Mohammadzadeh et al. (2005) propose "template guided association," aimed at XML. The DBS approach is similar: the templates are the schemata built from pattern proplets. In addition to a highly effective primary key, DBS provides the option to search for continuation values, morphosyntactic categories, base forms, matching inferences, memorable outcomes, n-grams, frequencies, and so on.

The conceptual backbone of DBS is storing proplets in the order of their arrival[6] in combination with the "no-change" rule of a content-addressable memory. Because the processing of content never modifies what is stored in memory, there is a strict separation between the storage of content and its processing. The storage operations, like the inferences, always write to the *now*

---

[5] Schegloff's examples are interesting and carefully analyzed, but a computational implementation was not one of Schegloff's goals. The observable facts are well documented, but the difficulties of interpretation or of choosing between several possible interpretations are pointed out repeatedly.

[6] In DBS, time is represented solely by the proplets' relative order of arrival. There are three possibilities: proplet A is earlier than than proplet B, proplet A is later than proplet B, or proplets A and B are simultaneous. This order is reflected by the prn values of the proplets.

front, using the same, simple, transparent procedure (4.1.1) to ensure historical accuracy and to conform to the structure of a content-addressable memory.

For correcting stored content, a comment is written to the *now front*, like a diary entry noting a change in temperature (cf. 4.4.2 for an analogous example). The comment refers to the content by means of addresses, providing instant access. When subactivation lights up a content, any and all addresses pointing at it are activated as well. When subactivation lights up an address, the original and all the other addresses pointing at it are also subactivated. In short, originals and their addresses are systematically *co-subactivated* in DBS.

12.2 RMD Corpus

While the core values, the semantic relations, and the levels of abstraction are agent-internal constructs of cognition, the language data are agent-external objects, collected, for example, as a corpus. Like any contemporary linguistic theory, DBS requires corpora to obtain frequency distributions of various kinds in a standardized framework. For example, when expanding automatic word form recognition, recognition rates will improve best if the most frequent word forms are integrated into the software first, and similarly for parsing syntactic-semantic constructions, and so on.

The frequency information should be obtained from a standardized RMD corpus, i.e., a Reference Monitor corpus structured into Domains. The reference corpus consists of a subcorpus for everyday language, complemented by subcorpora for different domains such as law, medicine, physics, sport, politics (cf. von der Grün 1998), including fiction, e.g., movie scripts. Their sizes may be determined by methods evolved from those used for the Brown corpus (Kučera and Francis 1967, Francis and Kučera 1982).

The reference corpus is continued with monitor corpora following every year (Sinclair 1991, p. 24–26). The annual monitor corpora resemble the reference corpus in every way: overall size, choice of domains, domain sizes, etc. The reference corpus and the monitor corpora use texts from a carefully selected set of *renewable* language data: newspapers for everyday language, established journals for specific domains, and a selection of fiction which appeared in the year in question.

Most of the corpus building and analysis may be done completely automatically. This holds (i) for the collecting of texts for the monitor corpora once the initial set of sources has been settled on, (ii) for the statistical analysis once a useful routine as been established. and (iii) for automatic word form recognition as well as syntactic-semantic parsing. Such automatic corpus processing replaces markup by hand, ensuring the quality of standardization necessary for meaningful comparisons, and saving the labor of instructing and the cost of remunerating large groups of markup personnel.[7]

A succession of monitor corpora allows a detailed view of how the language and the culture are developing, in different domains and over many decades. Statistical analysis will show, for example, how politics and natural disasters cause a temporary frequency increase of certain words in certain domains.

A carefully built long-term RDM corpus is in the interest of the whole language community and should be entrusted to the care of a national academy. This would secure the necessary long-term funding, though much of the cost could be recovered from commercial use of the continuously upgraded RDM corpus of the natural language in question (Sect. 12.6).

In DBS, the routine of analyzing a corpus begins with running the corpus through automatic word form recognition.[8] The result is a set of proplets called a *content* and stored in the content-addressable database of a Word Bank. Next, semantic relations are established between proplets by means of syntactic-semantic parsing. Finally, LA-think, inferences, and LA-speak are added.

Just as there is no limit to the amount of content stored in a Word Bank, at least in principle, there is no limit to the amount of information that can be added to the content. The information is integrated as a system of footnotes and subfootnotes. The "footnotes" are realized as interpreted pointers to other contents in the Word Bank. This does not increase the number of proplets in the Word Bank, only the number of addresses connecting them.

The user may query the Word Bank content in natural language (provided that the language software is available). Once an LA-think and an LA-speak grammar have been added, the answers may be in the natural language of the query. This method, though developed with carefully constructed input sentences, may eventually be applied to free text such as pages in the Internet. One benefit would be a quality of recall and precision unachievable by a statistical approach (cf. FoCL'99, Sects. 15.4, 15.5) or by manual markup.

12.3 Evolution

A computational model of natural language communication, defined at a level of abstraction which applies to natural and artificial agents alike, need not necessarily include the dimension of evolution. Instead, the software machine could be built as a purely "synchronic" framework of computational function.

---

[7] See Sect. 8.5 for the use of a corpus for the purpose of search space reduction in DBS.

[8] The usual preprocessing assumed.

Nevertheless, it would shed doubt on DBS as a theory if it turned out to be incompatible with the mechanisms of evolution. By the same token, if DBS can be shown to be analogous to evolution in some relevant process, this may increase interest in DBS for aspects other than synchronic performance alone.

The DBS analysis of learning[9] resembles evolution in that it starts from fixed behavior patterns (FAPs) and reconstructs learning by disassembling the parts and recombining them into adaptive behavior. For such a metamorphosis to be automatic, the steps from one state to the other must (i) be small enough to be driven by the interaction with the environment, (ii) be suitable for both "aggregate states," (iii) and evolve in a meaningful time-linear sequence, parallel branches not excluded. The software would be a kind of genetic algorithm, from which much could be learned for applications.

Another question raised by evolution is whether the language component evolved from (i) a nonlanguage component or from (ii) a non-communication component. This is in part a terminological question. The two dichotomies are language and nonlanguage and communication and non-communication.

For example, a chameleon picking up an insect with its long tongue uses a non-communication as well as a nonlanguage ability. Yet, like all living beings, chameleons also have an ability to communicate with their conspecifics.[10] As far as we know, the chameleon kind of communication does not satisfy criteria 3 and 4 for being a natural language listed in 2.2.3. Thus, the chameleon uses nonlanguage communication abilities, for example, by changing the color of its skin to attract a mate (Karsten et al. 2009).

The two dichotomies may be used to construct two hypotheses about the evolution of natural language, shown graphically as follows:

12.3.1 CONSECUTIVE VS. CONCURRENT HYPOTHESIS

a. Consecutive hypothesis:
language ability evolves from non–language abilities

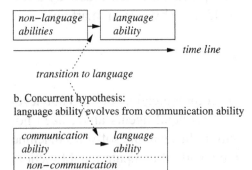

b. Concurrent hypothesis:
language ability evolves from communication ability

According to the consecutive hypothesis, the language ability evolved from nonlanguage abilities. Theoretically, these could be divided into nonlanguage communication and nonlanguage non-communication abilities. However, because the consecutive hypothesis does not provide for a communication component, language seems to evolve from non-communication abilities, which is wrong.

The concurrent hypothesis, in contrast, assumes that the communication and the non-communication abilities co-evolved from the outset. This provides the two abilities with the same amount of time to evolve. Thus, when language finally arrives as the last stage of the communication ability, it can rely on many nonlanguage communication abilities evolved earlier. It can also rely on a continuous, systematic interaction between the communication and the non-communication (context) components.

Communication ability is necessary to all living beings at least to the degree that reproduction is ensured. In evolution, reproduction is more important than the survival of individual members because reproduction serves the survival of the species as a whole. Communication is at the heart of evolution because reproduction is the motor driving evolution, and communication with conspecifics is at the heart of reproduction.

The concurrent hypothesis is in concord with modern work in ethology (M. D. Hauser 1996). It suggests that the communication and the non-communication abilities of an animal not only evolve in parallel, but also have matching degrees of development. Accordingly, the cognitive abilities of a squirrel to communicate with other squirrels, for example, are just as evolved as its abilities for locomotion and behavior control.[11]

Furthermore, because even nonlanguage communication refers,[12] the communication and the non-communication abilities do not just grow in parallel, but are continuously in close interaction. In DBS, this interaction is modeled as a collaboration between the language and the context components. Natural language signs and contextual data interact in a software-mechanical way during interpretation in the hear mode and production in the speak mode. In short, designing the DBS robot in an evolutionary manner is another way of broadening the empirical base.

The above considerations regarding the evolution of natural language have a direct effect on the software design of DBS: if the communication and the

---

[9] Sects. 6.1–6.3.

[10] In addition to communication between conspecifics there is also the communication between different species, for example, between a plant and an insect for pollination (symbiosis).

[11] It would be thrilling to know what the little critters all have to communicate about.

[12] For example, a warning call makes reference to a predator.

noncommunication components are in such close functional interaction and at such equal levels of evolutionary development, the simplest approach is to program the two components essentially alike while paying close attention to their interaction with each other and with their external interfaces. Designing, using, and reusing software constructs which are as uniform and simple as possible is also regarded as good programming practice.

12.4 Semantics

Is DBS really a semantics? Let us answer this question by a point by point comparison with the reigning queen of semantics, Symbolic Logic.

The most important property common to Symbolic Logic and Database Semantics is that they are both Aristotelian. At the core of the Aristotelian approach is the use of only two basic semantic relations of structure, namely (i) coordination and (ii) functor-argument.

In Symbolic Logic, this fundamental insight is realized by (i) propositional calculus for extrapropositional coordination and (ii) predicate calculus for intrapropositional functor-argument. These may be extended into (iii) extrapropositional functor-argument [13]and (iv) intrapropositional coordination.[14]

Extrapropositional coordination is represented in Symbolic Log by expressions like $((p \wedge q) \vee r)$. They are composed by the syntactic-semantic rules of propositional calculus (cf. FoCL'99, 19.3.2).

In DBS, the logical connectives are introduced by function words, lexically analyzed by automatic word form recognition. Called conjunctions, they carry their semantic contribution as a value inside their proplet representation. In DBS, the extrapropositional coordination corresponding to $f(a) \wedge f'(a') \wedge f''(a'')$ is shown by example in 3.2.1 and abstractly as the schema 3.2.6.

Intrapropositional functor-arguments are represented in Symbolic Logic by expressions like $f(a,b)$, where f is a functor and a and b are arguments. These

---

[13] However, the Donkey sentence 11.5.5 shows that the extension to subclauses is not always possible. Though a lower level defect, it has been regarded as serious enough to spawn a massive body of literature trying to repair it – within, or almost within, the tradition of predicate calculus.

[14] Extending propositional calculus from extrapropositional to intrapropositional coordination also creates a problem for Symbolic Logic. It arises with examples like **All the students gathered**; **John and Mary are a happy couple**; **Suzy mixed the flower, the sugar, and the eggs**; etc., which intuitively suggest a collective, and not a distributive, reading (see Zweig 2008 for a overview; see also Hausser 1974; Kempson et al. 1981).

The "mix" problem does not arise in extrapropositional coordination, just as the "donkey" problem does not occur in intrapropositional functor-argument. Within Symbolic Logic, the "mix problem" can be solved by proposing new quantifiers and/or connectives, while the "donkey" problem is caused by failing scope and cannot be solved without uprooting the quantifier-based syntax and the bracketing structure of predicate calculus. The DBS solution is based on the use of addresses (11.5.7).

may be viewed as rudimentary parts of speech: f is like a two-place verb, while a and b are like nouns serving as subject and object.[15]

The corresponding constructs in DBS are features, defined as attribute-value pairs, using noun, verb, and adj as the core attributes of proplets. By distinguishing between the attribute and the value(s), there may be features such as [noun: dog] which are more differentiated and intuitive than the individual constants a_1, a_2, a_2, ... of Symbolic Logic, without any loss of generality.[16] In DBS, a functor-argument like f(a,b) of predicate calculus is shown by example in 3.2.5 and represented abstractly as the schema 3.2.6.

In any semantics, there are two kinds of meaningful elements, (i) *logical* and (ii) *contingent*. In predicate calculus, the logical elements are the connectives, the quantifiers, the variables, and a certain bracket syntax, while the contingent elements are letters for constants. In Database Semantics, the logical elements are the attributes and the variables in the proplets, while the non-variable values, i.e., the symbols, indexicals, and names, are contingent.

The traits common to Symbolic Logic and DBS may be increased further by taking the liberty to reinterpret the semantic interpretation of a sign-oriented approach as the *hear mode* of an agent-oriented approach.[17] In this way, the sign-oriented approach may be taken to cover one of the three steps of the natural language communication cycle.

This holds especially for Montague (1974), who shows in PTQ how to semantically interpret surfaces of English by translating them into formulas of predicate calculus. The translation is quasi-mechanical in that the reader can go mentally through the rule applications and the lambda reductions as if going through the steps of a proof. Montague formalizes the translation mechanism as a Categorial Grammar with (i) a cleverly structured set-theoretical interpretation and (ii) the reduction mechanism of typed lambda calculus (lambda reduction).[18]

From the viewpoint of this agent-oriented reinterpretation, DBS may be seen as completing Montague grammar in two ways. One completion is the extension of Montague grammar to the full cycle of natural natural language communication by adding the think and the speak mode. The other completion is

---

[15] Using typed lambda calculus, Montague (1974) worked hard to formalize functors and arguments set-theoretically as semantic types with corresponding syntactic categories, fitting into the rule schemata of Categorial Grammar. Lambda reduction in a typed lambda calculus may be viewed as a souped-up version of the categorial canceling rules.

[16] If needed, a feature may be represented abstractly as a pattern with the attribute and the values represented by variables. Cf. NLC'06, Sect. 4.1.

[17] This reinterpretation is the founding assumption of SCG'84.

replacing Montague's quasi-mechanical interpretation method with the computational automation provided by efficiently running software.

These two completions, however, have necessitated many solutions different from Montague grammar in particular and Symbolic Logic in general. For example, predicate calculus uses the logical elements to construct a syntactic exoskeleton, while DBS integrates the logical and the contingent elements into flat feature structures with ordered attributes (proplets) which code inter-proplet relations solely by addresses, and are therefore order-free (3.2.4).

The difference between DBS and predicate calculus, in particular regarding the role of quantifiers, may be shown in more detail with the following example from Montague (1974, PTQ):

12.4.1 PREDICATE CALCULUS ANALYSIS OF Every man loves a woman.

reading 1: $\forall x[\text{man}'(x) \rightarrow \exists y[\text{woman}'(y) \,\&\, \text{love}'(x, y)]]$

reading 2: $\exists y[\text{woman}'(y) \,\&\, [\forall x[\text{man}'(x) \rightarrow \text{love}'(x, y)]]$

This standard[19] analysis of predicate calculus treats the English surface as syntactically ambiguous. The reason is a scope ambiguity suggested by the creaking hinges of its quasi-mechanical exoskeleton. The contingent elements are man′, woman′, and love′.

In DBS the ambiguity alleged in 12.4.1 is not syntactic-semantic, but at best pragmatic.[20] The content of the sentence is shown as a set of proplets:

12.4.2 DBS ANALYSIS OF Every man loves a woman.

$$
\begin{bmatrix} \text{noun: man} \\ \text{cat: snp} \\ \text{sem: exh pl} \\ \text{fnc: love} \\ \text{prn: 23} \end{bmatrix}
\begin{bmatrix} \text{verb: love} \\ \text{cat: decl} \\ \text{sem: pres} \\ \text{arg: man woman} \\ \text{prn: 23} \end{bmatrix}
\begin{bmatrix} \text{woman} \\ \text{cat: snp} \\ \text{sem: indef sg} \\ \text{fnc: love} \\ \text{prn: 23} \end{bmatrix}
$$

In DBS, the logical quantifier $\forall x$ is represented alternatively by the **sem** values **exh pl** (exhaustive plural) of the *man* proplet, while the quantifier $\exists y$ is

---

[18] When writing SCG'84 we had to learn the hard way that a typed calculus is not very practical. Even worse, the small fragment we had managed to define in SCG'84 turned out to be unprogrammable. The solution, published as NEWCAT'86, is a strictly time-linear derivation order.

[19] Montague's main contribution is the quasi-mechanical translation of the English surface into formulas of predicate calculus. The number of grammatical constructions is rather small, and motivated in part by concerns of analytic philosophy, such as *de dicto* and *de re*.

[20] The distinction between syntactic, semantic, and pragmatic ambiguities is explained in FoCL'99, Sect. 12.5. Only syntactic ambiguities are of complexity-theoretic relevance. The syntactic ambiguity of the example 12.4.1 seems to originate mostly in classes on predicate calculus.

represented by the **sem** values **indef sg** (indef singular) of the *woman* proplet.[21] The verb's **cat** value **decl** (for declarative) and its **sem** value **pres** (for
present tense) complete the picture. The proplets form an order-free set, and
are held together by a common **prn** value, here **23**.

In summary, predicate calculus uses (i) variables within formulas representing content and (ii) quantifiers (a) to bind the variables *horizontally* and (b)
to replicate determiners at the same time. DBS, in contrast, (i) replaces the
binding function of quantifiers by a **prn** value, (ii) employs variables solely
for a *vertical* binding between pattern and content proplets, and (iii) codes the
determiner function of quantifiers as values of the **cat** and **sem** attributes.

With the elimination of quantifiers, DBS can restrict the use of all kinds
of variables to the definition of *pattern* proplets. The pattern proplets are
combined into schemata and are used for matching with content by the LA-
grammar rules and for retrieval (activation) in the content-addressable memory of a Word Bank. Using restricted variables as proplet values is a simple,
powerful method of under-specification,[22] with many uses in DBS.

Up to this point, DBS may be regarded as a modern reconstruction of predicate calculus, though with special emphasis on automatic natural language
interpretation, processing of content, and natural language production – rather
than on truth, satisfiability, consistency, and completeness. The latter are important, but they are neither intended nor sufficient for building a model of
natural language communication.

Finally, let us consider meaning, which is what semantics is all about. Logicians take great care to give their formulas a semantic interpretation. It is a
conceptual construction in which "the world" is defined in principle as a set-
theoretical model structure, and the surfaces of the variables and constants of
a formula are defined by hand to refer to these set-theoretical constructs using
a metalanguage. The purpose is inferencing and theorem proving.

Instead of treating concepts in terms of metalanguage definitions, DBS treats
concepts as the basic *procedures* of the agent's recognition and action. Declarative representations of these procedures are reused[23] as the meanings of language, with the type/token relation serving in pattern matching (4.3.3). This
procedural approach to concepts and meanings demands the switch from a

---

[21] The values **exh**, **sel**, **def**, **indef**, **sg**, and pl are defined set-theoretically in NLC'06, 6.2.9.

[22] Restricted variables are inherently open-ended. When used for matching, variables may be bound
tentatively to values not in their restriction set. Also, when new selectional constellations (Chap. 8)
have been found, these must be added to the relevant restriction sets.

[23] Cf. Roy 2005 for a similar approach.

sign-oriented [-sense, -constructive] ontology to an agent-oriented [+sense, +constructive] ontology (cf. FoCL'99, Sect. 20.4).[24]

12.5 Autonomous Control

An agent-oriented approach requires an autonomous control for sensible behavior, including natural language production. In DBS, autonomous control is driven by the principle of balance. A state of balance is an absolute like truth, and like truth it provides the fix point necessary for a system of semantic interpretation. But while truth is (at least[25]) bipolar, balance is monopolar. This may be the reason why balance, unlike truth, is a dynamic principle, suitable for driving the cognitive agent trying to survive, and to survive comfortably, in a constantly changing world, short-, mid-, and long-term.

In DBS, the agent's search for balance is implemented as a set of inferences. An inference is triggered by an activated content matching the antecedent. After binding the variables of the antecedent to constants in the input, e.g., core values, the consequent of the inference derives a blueprint for action as output. The output is written to the *now front* of the agent's Word Bank.

This new approach to inferencing is based on the belated insight that the database schema of a Word Bank, designed before FoCL'99, is in fact a *content-addressable* memory.[26] Content-addressable memories happen to be the most efficient for content written once and never changed. By structuring the memory like *sediment*, written once and never changed, all processing for real-time behavior control is restricted to the *now front*.

Stored content which never changes makes practically[27] no processing demands on the system. At the same time, the personal history contained in an agent's static sediment provides an individual notion of what is true and what is right. The agent's current state is defined by the most recent data, stored last (rightmost) in the token lines.

The agent's overall moment to moment behavior may be viewed as regulated by of a basket of weights which represent the current options for maintaining

---

[24] The binary feature notation without attributes used here, e.g., [+constructive], resembles the "feature bundles" of Chomsky and Halle (1968).

[25] Cf. FoCL, Sect. 20.5.

[26] Originally, the Word Bank had been naively derived from *classic* (i.e., record-based) *network databases* as described in Elmasri and Navathe (1989). Driven by functional concerns, our interest was focused primarily on a running program, working as intended. Over the years, the implementation of a Word Bank has been explored in a sequence of at least four projects at the CLUE.

[27] With the exception of occasional cleanups, cf. Sect. 5.6.

or regaining balance.[28] Which blueprint for action is selected for realization is determined by continuously recalculating the weights associated with the agent's current needs and the consequences of available options. The calculation is based on a cost-benefit analysis, computed in real-time.

Fine-tuning the DBS system to simulate the balance of a robot in a terrain, i.e., in a co-designed changing environment, requires an actual robot. After all, without the robot's external and internal interfaces we would have to recreate every nook and cranny of the changing environment by hand (as in Model Theory, cf. 4.3.1). This would violate a basic principle of nouvelle A.I., namely that *The world is its own best model* (Brooks 1989).

To manage the massive, multiple, parallel search required for operating a DBS robot in real time, retrieval must be based on efficient database operations and provide highly differentiated recall.[29] This task may and must be solved theoretically, i.e., without any need for actual robot hardware. The following aspects may be distinguished: (i) the organization of competing retrieval tasks at any given moment and (ii) the quality of the search mechanism itself.

For the synchronization of competing retrieval tasks, DBS uses the time-linear derivation order (cf. Herlihy and Shavit, in press, for related issues). All parallel derivation strands apply their current step, e.g., a rule application, simultaneously. An example of time-linear derivation strands running in parallel is the simultaneous operation of (i) the hear mode, (ii) subactivation, (iii) intersection, and (iv) inferencing.

In a step, each strand produces a set of retrieval tasks, specified by the DBS rules and database operations to be applied. These jobs may be executed in parallel or sequentially in some suitable order, depending on the hardware and the operating system. This approach to organizing parallel operations is not only efficient but also transparent – which is essential for debugging and optimization of the DBS robot.

The other aspect of optimal retrieval is the quality of the individual search operations. In DBS, it (i) depends on the speed of the retrieval mechanism and (ii) must have the expressive power needed for the kind of queries to be expected. For speed, DBS uses the schema of a content-addressable database (Word Bank) and the use of pointers. For expressive power, DBS utilizes the

---

[28] As shown in Chaps. 5 and 6, this is based in part on rule-governed and goal-governed inferencing, fixed behavior, and trial and error.

[29] In the hear mode, retrieval is used to determine the correct token line for storing proplets at the *now front*. In the think mode, retrieval must activate a successor proplet somewhere in the Word Bank, using an address. In the speak mode, the content to be realized must be mapped into a sequence of language surfaces, written to the *now front*, which again requires finding the proper token line. Activated content must find inferences with a matching antecedent.

semantic relations of coordination and functor-argument, defined at (a) the content and (b) the schema (rule) levels. Introduced to model the cycle of natural language communication, the semantic relations provide the structural basis also for subactivation, intersection, and inferencing.

Because the content in a Word Bank is structured by the semantic relations of natural language, the system can respond in kind, i.e., it can be as specific or general as formulated in the query or any other search request. This ability to respond to language questions with language answers, developed for natural language dialogue with a talking robot, may also be used for more conventional applications. For example, a database structured as a Word Bank, sitting on a standard computer in some geographically remote warehouse, may be queried, and may answer, in natural language.

12.6 Applications

The DBS approach to practical (commercial) applications of natural language processing is based on solving the most important theoretical question first: *How does the mechanism of natural language communication work?*

To protect against accidentally neglecting some crucial interface, component, or ability, the overall design of a DBS robot aims at functional completeness. By modeling all essential structural aspects of natural language communication by humans it is hoped that there will be no application-motivated requests which cannot be satisfied.

If a functional framework works properly at all levels of abstraction, though with small (and highly relevant) data coverage only, then all that remains to be done is to increase the data coverage. For natural language communication, this is a mammoth project, though nothing compared to projects in physics (CERN) or biology (human genome project), for example.

Extending the data coverage as a form of upscaling has immediate consequences on commercial applications using the system for their natural language processing needs. Take for example LA-morph, the automatic word form recognition software, running with a certain natural language of choice.

The data coverage of such an instance of LA-morph may be extended by adding to the lexicon and by optimizing the allo- and combi-rules for the natural language at hand. This broadens the base for syntactic-semantic analysis and inferencing. It also provides practical applications with better results for retrieval based on content words.

A second area for completing data coverage is extending the syntactic-semantic analysis. When applied to a new (i.e., previously unanalyzed) natural language, the LA-hear parser will at first handle only a few constructions. As the language is being studied, more and more constructions (like infinitives, prepositional phrases, relative clauses, etc.) are added to the grammar, tested, and revised. When the LA-hear parser encounters input it cannot yet handle, the passage may be traversed at a lower level of detail until proper parsing can resume (robustness). For this, LA-grammar is especially suitable because it computes possible continuations in a time-linear derivation order.

Expanding syntactic-semantic parsing in the agent's hear mode is more demanding than automatic word form recognition. This effort should not go unrewarded from the application side, however. The coding of functor-argument and coordination extends recall and precision from lexically analyzed word forms to phrases and clauses, and from there to sentences, paragraphs, and text. Technically, this amounts to an extension from matching lexically analyzed content words stored within token lines in the Word Bank, to matching semantic relations between content words defined across token lines.[30]

The think mode is a third area for extending the data coverage. The agent's think mode combines two mechanisms, LA-think and inferencing. The basic mechanism of LA-think is *selective activation* by navigating along the semantic relations in a Word Bank.[31] The navigation is used to activate and report self-contained content.

Inferences are used for deriving the different perspectives of the speaker and the hearer on content,[32] and to compute blueprints for action, including language action. Together with current and stored data, LA-think and inferencing constitute the agent's autonomous control, which has many practical applications, with and without language.

Finally, consider LA-speak. It takes content as input and produces corresponding surfaces as output. If the content has already been serialized by the navigation along the semantic relations in the Word Bank, the task of LA-speak is confined to adjusting to the word order of the language and to providing proper lexicalization with proper perspective (e.g., tense) and proper morphosyntactic adjustments (e.g., agreement).

---

[30] In addition, the user may load the proplets in a proprietary database with additional attributes and values as needed for the application. One such application of LA-morph and LA-hear is speech recognition; it could well benefit from the search space reduction resulting from an LA-hear parser computing possible continuations (Sect. 2.4).

[31] A Word Bank may be viewed as a syntactic-semantic network. For some questions and results of linguistic networks, see Liu (2011), Solé et al. (2010), Sowa (1987/1992), Brachman (1979), and others.

[32] The interaction between LA-think, LA-speak, and inferencing is shown in Chap. 10 with an example of dialogue.

This work will not go unrewarded from the application side either. The obvious application is *query answering* in natural language. Thereby the LA-speak part is only the tip of the iceberg. Prior to answering, the query is converted automatically into several schemata which are used to subactivate corresponding contents. These data are processed into the content for the query answer by means of intersection and inferencing. Once the resulting answer content has been derived, it is passed to LA-speak for realization as an unanalyzed external surface.

While specific applications may benefit selectively from the nurturing of a particular component, all applications will benefit simultaneously from a methodical upscaling of the DBS robot as a whole. An application which does not require certain abilities may be run with a DBS version in which they have been switched off.[33]

The systematic, theory-driven upscaling of a talking robot is of general interest for the following reasons. First, it provides the opportunity to ensure compatibility between the system's components in a declarative manner.[34] Second, the neighboring sciences, for example, psychology, ethology, neurology, philosophy, etc., may use the computational model to test some of their own issues, which may in turn contribute to the long-term effort of upscaling the talking robot.[35] Third, by making regular version updates available to the public, progress in pure research may quasi automatically improve the language processing of participating applications.

The orderly transfer from a continuously improving DBS system to commercial applications of human-machine communication may be illustrated by the following vision. Every year, when the current monitor corpus (Sect. 12.2) has been put through the automatic software grinder of word form recognition, parsing, frequency analysis, and comparison with preceding monitor corpora, the results are used for a software version with improved data coverage.

By making new versions available to paying subscribers for their natural language processing needs, all or most of the research costs may be recovered. For this to work long-term, a new release must not require any labor from the subscriber (e.g., additional personnel training), except for the routine installation. Also, each new version must enhance service directly and noticeably, so that subscribers are attracted and kept in sufficient numbers.

---

[33] For example, a dialogue system over the phone may omit the ability to read.

[34] In addition, typing (in the sense of computer science) could be used in DBS. It is not really needed, however, because of the simplicity and formal uniformity of proplets and the associated interfaces and algorithm.

[35] For example, a system of control implemented in analogy to the sympaticus/parasympaticus nerve network.

Improvements from one version to the next may be achieved rather easily because there are large fields of empirical data which merely need to be "harvested." The software machine for the systematic collection, analysis, and interpretation of the language data is the DBS robot, originally designed to model the mechanism of natural language communication.

For example, when applied to a new language, the DBS robot's off-the-shelf components for the lexicon, automatic word form recognition, syntactic-semantic parsing, and so on, hold no language-dependent data. As a new language is being analyzed, words are added to the robot's lexicon component, just as compositional structures are added to the LA-Morph, LA-hear, LA-think, and LA-speak grammars in the robot's rule component. Also, culture-dependent content may be added to the Word Bank.

Storing the analysis of a natural language directly in the DBS robot makes the analysis available right away for computational testing by the scientists and for computational applications by the users. This works not only for the hear mode, as in testing on a corpus, but for the full cycle of natural language communication. The testing is designed (i) to automatically enhance the robots performance by learning, and (ii) to provide the scientists with insights for improving the robot's learning abilities.

For long-term linguistic research, there is no lack of renewable language data, namely (i) the natural changes year to year within the domains of a given language and (ii) a wide, constantly extending range of applications in human-machine communication. In addition, there is (iii) the great number of natural languages not yet charted, or not yet charted completely (including English, in any theory). The harvesting of each of these kinds of data will be of interest to its own group of users.

Charting a new natural language is a standard procedure, but it has to deal with relatively large amounts of data. As more and more languages are analyzed, however, charting is accelerated because software constructs may be reused, based on similarities in lexicalization, in productive syntactic-semantic structures, in collocations, constructions, and idioms, and in inferencing. To better support day-to-day research,[36] these standardized software constructs and their declarative specifications may be stored in system libraries, organized for families of languages.

---

[36] For example, in work on typology or on expanding a given language to new constructions.

List of Examples, Definitions, Figures, and Tables

In this text, linguistic examples, figures, definitions, and tables follow headings with a three-part label referring to the chapter, the section, and the item number. For convenience, these numbered headings, altogether 300, are listed below:

Part II: The Coding of Content

7. Compositional Semantics

Part III: Final Chapter

12. Conclusion

Bibliography

Ágel, V., L. Eichinger, H.-W. Eroms, P. Hellwig, H.-J. Heringer, and H. Lobin (eds.) (2006) *Dependenz und Valenz: Ein Internationales Handbuch der zeitgenoessischen Forschung,* Vol. 2, Berlin: De Gruyter

Aho, A. V. and J. D. Ullman (1977) *Principles of Compiler Design,* Reading, MA: Addison-Wesley

AIJ'01 = Hausser, R. (2001) "Database Semantics for Natural Language," *Artificial Intelligence,* Vol. 130.1:27–74

Ajdukiewicz, K. (1935) "Die syntaktische Konnexität," *Studia Philosophica,* Vol. 1:1–27

Anderson, J. R. (1983) "A Spreading Activation Theory of Memory," *Journal of Verbal Learning and Verbal Behavior,* Vol. 22:261–295

Arena, P. and L. Patanè (eds.) (2009) *Spatial Temporal Patterns for Action-Oriented Perception in Roving Robots,* Cognitive Systems Monographs, Springer

Bachman, C. W. (1973) "The Programmer as Navigator," 1973 ACM Turing Award Lecture, *Comm. ACM,* Vol. 16:11.653–658

Bar-Hillel, Y. (1953) "Some Linguistic Problems Connected with Machine Translation," *Philosophy of Science,* Vol. 20:217–225

Bar-Hillel, Y. (1964) *Language and Information. Selected Essays on Their Theory and Application,* Reading, Mass.: Addison-Wesley

Barsalou, L. (1999) "Perceptual Symbol Systems," *Behavioral and Brain Sciences,* Vol. 22:577–660

Barthes, R. (1986) *The Rustle of Language,* New York: Hill and Wang

Barwise, J. and J. Perry (1983) *Situations and Attitudes,* Cambridge, MA: MIT Press

Bauer, M. (2011) *An Automatic Word Form Recognition System for English in JSLIM,* M.A. thesis, CLUE (Computational Linguistics U. Erlangen)

Beauvoir, S. de (1960) *La force de l'âge,* Paris: Gallimard

ben-Zineb, R. (2010) *Implementierung eines automatischen Wortformerkennungssystems für das Arabische in JSLIM,* M.A. thesis, CLUE

Bernard. C. (1865) *Introduction à l'étude de la médecine expérimentale,* first English translation by Henry Copley Greene, published by Macmillan, 1927; reprinted 1949

Berwick, R. C. and A.S. Weinberg (1984) *The Grammatical Basis of Linguistic Performance: Language Use and Acquisition*, Cambridge, Mass.: MIT Press

Beutel, B. (2009) *Malaga, Grammar Development Environment for Natural Languages*, http://home.arcor.de/bjoern-beutel/malaga/

Beynon-Davies, P. (2003) *Database Systems*, Basingstoke: Palgrave Macmillan

Biederman, I. (1987) "Recognition-by-Components: a Theory of Human Image Understanding," *Psychological Review*, Vol. 94:115–147

Boas, F. (1911) *Handbook of American Indian languages (Vol. 1)*, Bureau of American Ethnology, Bulletin 40. Washington: Government Print Office (Smithsonian Institution, Bureau of American Ethnology)

Boden, D. and D. Zimmerman (eds.) (1984) *Talk and Social Structure*, Berkeley: U. of California Press

Borsley, R. and K. Borjars (eds.) (in press) *Non-Transformational Syntax*, Cambridge: Wiley-Blackwell

Bosch, P. (1986) "Pronouns under Control? A Reply to Lilliane Tasmosvki and Paul Verluyten," *Journal of Semantics*, Vol. 5:65–78

Brachman, R. J. (1979) "On the Epistemological Status of Semantic Networks," in N. Findler (ed.), pp. 3–50; Reprinted in Brachman and Levesque (eds.) 1985

Brachman, R. J. and Levesque (eds.) (1985) *Readings in Knowledge Representation*, Los Altos, CA: Morgan Kaufmann

Braitenberg, V. (1984) *Vehicles: Experiments in Synthetic Psychology*, Cambridge, MA: MIT Press

Brinton, L. J. and E. Closs Traugott (2005) *Lexicalization and Language Change*, CUP

Brooks, R. (1985) "A Robust Layered Control System for a Mobile Robot," Cambridge, MA: *MIT AI Lab Memo 864*, pp. 227–270

Brooks, R. (1990) "Elephants Don't Play Chess," MIT Artificial Intelligence Laboratory, *Robotics and Autonomous Systems,* Vol. 6:3–15

Brooks, R. (1991) "Intelligence without Representation," *Artificial Intelligence*, Vol. 47:139–160

Buszkowski, W. (1988) "Gaifman's Theorem on Categorial Grammars Revisited," *Studia Logica*, Vol. 47.1:23–33

Campbell, N. A. (1996) *Biology, 4th ed.*, New York: Benjamin Cummings

Carbonell, J. G. and R. Joseph (1986) "FrameKit+: a Knowledge Representation System," Carnegie Mellon U., Department of Computer Science

Carpenter, B. (1992) *The Logic of Typed Feature Structures*, Cambridge: CUP

CELEX2(1996) Olac Record, http://olac.ldc.upenn.edu/item/oai:www.ldc.upenn.edu:LDC96L14

Chen, P. (1976) "The Entity-Relationship Model – Toward a Unified View of Data," *ACM Transactions on Database Systems*, pp. 9–36 ACM-Press

Chisvin, L. and R. J. Duckworth (1992) "Content-Addressable and Associative Memory," pp. 159–235, in M.C. Yovits (ed.)

Chomsky, N. and M. Halle (1968) *The Sound Pattern of English*, New York: Harper Row

Cole, R. (ed.) (1997) *Survey of the State of the Art in Human Language Technology*, Edinburgh: Edinburgh U. Press

Comrie, B. (1986) "Reflections on Subject and Object Control," *Journal of Semantics*, Vol. 4:47–65, Oxford: OUP

Corbett, A. T., J. R. Anderson, and E. J. Patterson (1988) "Problem Compilation and Tutoring Flexibility in the LISP Tutor," pp. 423–429, Montréal: ITS-88

Cormen, T. H., C. E. Leiserson, R.L. Rivest, and C. Stein (2009) *Introduction to Algorithms, 3rd ed.*, Cambridge, MA: MIT Press

Croft, W. (2000) "Parts of Speech as Language Universals and as Language-Particular Categories," in Vogel and Comrie 2000, pp. 65–102

Cuvier, G. (1817/2009) *Le règne animal; distribué d'après son organisation; pour servir de base à l'histoire naturelle des animaux et d'introduction à l'anatomie comparé,* 4 Vol., Paris

Darwin, C. (1839/1989) *The Voyage of the Beagle*, Penguin Classics

Davidson, D. and G. Harman (eds.) (1972) *Semantics of Natural Language*, Dordrecht: D. Reidel

Dik, S. (1989) *The Theory of Functional Grammar (Part I: The Structure of the Clause)*, Berlin, New York: Mouton de Gruyter

Dik, S. (1997) *The Theory of Functional Grammar (Part II: Complex and Derived Constructions)*, Berlin, New York: Mouton de Gruyter

Dominey, P. F. and F. Warneken (in press) "The Basis of Shared Intentions in Human and Robot Cognition," *New Ideas in Psychology*, Elsevier

Dorr, B. J. (1993) *Machine Translation: A View from the Lexicon*, Cambridge, MA: MIT Press

Dodt, E. and Y. Zotterman (1952) "The Discharge of Specific Cold Fibres at High Temperatures; the Paradoxical Cold," *Acta Physiol Scand.*, Vol. 26(4):358–365

Elmasri, R. and S. B. Navathe (1989) *Fundamentals of Database Systems, 5th ed. 2006*. Redwood City, CA: Benjamin-Cummings

Endsley, M. and D. J. Garland (eds.) (2000) *Situation Awareness and Measurement*, Mahwah, New Jersey: Lawrence

Endsley, M., B. Bolte, and D. G. Jones (2003) *Designing for Situation Awareness*, London and New York: Taylor and Francis

Fikes, R. E. and N. Nilsson (1971) "STRIPS: A New Approach to the Application of Theorem Proving to Problem Solving," *Artificial Intelligence*, Vol. 2:189–208

Fillmore, C., P. Kay, and C. O'Connor (1988) "Regularity and Idiomaticity in Grammatical Constructions: The Case of Let Alone," *Language*, Vol. 64: 501–38

Findler. N. (ed.) (1985) *Associative Networks*, Academic Press

Fischer, W. (2002) *Implementing Database Semantics as an RDMS* (in German), Studienarbeit am Institut für Informatik der Universität Erlangen-Nürnberg (Prof. Meyer-Wegener), published as CLUE-Arbeitsbericht 7 (2004), available at http://www.linguistik.uni-erlangen.de/clue/de/arbeiten/arbeitsberichte.html

FoCL'99 = Hausser, R. (1999) *Foundations of Computational Linguistics, Human-Computer Communication in Natural Language, 2nd ed. 2001*, Springer

Fowler, M. (2003) *UML Distilled: A Brief Guide to the Standard Object Modeling Language, 3rd ed.*, Addison-Wesley

Francis, W. N. and H. Kučera (1982) *Frequency Analysis of English Usage: Lexicon and Grammar*, Boston: Houghton Mifflin

Fredkin, E. (1960) "Trie Memory," *Commun. ACM*, Vol. 3.9:490–499

Frege, G. (1967) *Kleine Schriften*, I. Angelelli (ed.), Darmstadt: Wissenschaftliche Buchgesellschaft

Frege, G. (1879) *Begriffsschrift. Eine der arithmetischen nachgebildete Formelsprache des reinen Denkens*, Halle: L. Nebert

Gao, M. (2007) *An Automatic Word Form Recognition System for German in Java*, M.A. thesis, CLUE

Gazdar, G. (1982) "Phrase Structure Grammar," in P. Jacobson and G.K. Pullum (eds.)

Geach, P. (1969) "A Program for Syntax," in D. Davidson and G. Harman (eds.), pp. 483–497

Geurts, B. (2002) "Donkey Business," *Linguistics and Philosophy*, Vol. 25:129–156

Girstl, A. (2006) *Development of Allomorph-Rules for German Morphology Within the Framework of JSLIM*, B.A. thesis, CLUE, Universität Erlangen-Nürnberg'

Givón, T. (1990) *Syntax: A Functional Typological Introduction* (2 Volumes), Amsterdam: John Benjamins

Grice, P. (1957) "Meaning," *Philosophical Review*, Vol. 66:377–388

Grice, P. (1965) "Utterer's Meaning, Sentence Meaning, and Word Meaning," *Foundations of Language*, Vol. 4:1–18

Grice, P. (1969) "Utterer's Meaning and Intention," *Philosophical Review*, Vol. 78:147–77

Groenendijk, J. A. G., T. M. V. Janssen, and M. B. J. Stokhof (eds) (1980) *Formal Methods in the Study of Language*, U. of Amsterdam: Mathematical Center Tracts 135

Grün, A. von der (1998) *Wort-, Morphem- und Allomorphhäufigkeit in domänenspezifischen Korpora des Deutschen*, M.A. thesis, CLUE

Hajičová, E. (2000) "Dependency-based Underlying-Structure Tagging of a Very Large Czech Corpus," in: Special issue of TAL Journal, Grammaires de Dépendence/Dependency Grammars, pp. 57–78, Paris: Hermes

Halliday, M. A. K. and R. Hasan (1976) *Cohesion in English*, London: Longman

Handl, J. (2008) *Entwurf und Implementierung einer abstrakten Maschine für die oberflächenkompositionale inkrementelle Analyse natürlicher Sprache*, Diplom thesis, Department of Computer Science, U. of Erlangen-Nürnberg

Handl, J., B. Kabashi, T. Proisl, and C. Weber (2009) "JSLIM – Computational Morphology in the Framework of the SLIM Theory of Language," in C. Mahlow and M. Piotrowski (eds.)

Handl, J. and C. Weber (2010) "A Multilayered Declarative Approach to Cope with Morphotactics and Allomorphy in Derivational Morphology", in *Proceedings of the 7th Conference on Language Resources and Evaluation (LREC 2010)*, pp. 2253–2256, Valletta, Malta

Handl, J. (2011) *Some Formal Properties of DBS*, Ph.D. thesis, CLUE

Hauser, M. D. (1996) *The Evolution of Communication*. Cambridge, MA: MIT Press

Hausser, R. (1974) "Syntax and Semantics of Plural," *Papers from the Tenth CLS-Meeting*, Chicago: Chicago Linguistics Society

Hausser, R. (2003) "Reconstructing Propositional Calculus in Database Semantics," in H. Kangassalo et al. (eds.)

Hausser, R. (2007) "Comparing the Use of Feature Structures in Nativism and in Database Semantics," in H. Jaakkola et al. (eds.)

Hausser, R. (2008) "Center Fragments for Upscaling and Verification in Database Semantics," in Y. Kiyoki et al. (eds.)

Hausser, R. (2009) "Modeling Natural Language Communication in Database Semantics," in M. Kirchberg and S. Link (eds.), *Proceedings of the APCCM 2009*, Australian Computer Science Inc., CRPIT, Vol. 96:17–26

Hausser, R. (2010) "Language Production in Database Semantics," in PACLIC 24, *Proceedings of the 24th Pacific Asia Conference on Language, Information, and Computation*, 4–7 November 2010, Tohoku U., Sendai, Japan

Helfenbein, D. (2005) *Die Behandlung pronominaler Koreferenz in der Datenbanksemantik*, M.A. thesis, CLUE

Hengeveld, K. (1992) *Non-Verbal Predication: Theory, Typology, Diachrony*, Berlin, New York: Mouton de Gruyter

Herbst T., D. Heath, I. F. Roe, and D. Götz (2004) *A Valency Dictionary of English: A Corpus-Based Anaysis of the Complementation Patterns of English Verbs, Nouns, and Adjectives*. Berlin, New York: Mouton de Gruyter

Herlihy, M. and N. Shavit, in press, "On the Nature of Progress," ACM

Herrmann, T. (2003) "Kognitive Grundlagen der Sprachproduktion," in G. Rickheit, Th. Herrmann, and W. Deutsch (eds.), *Psycholinguistik-Psycholinguistics. Ein internationales Handbuch.* pp. 228–244, Berlin, New York: Mouton de Gruyter,

Huddleston R. and G. K. Pullum (2002) *The Cambridge Grammar of the English Language*, Cambridge: CUP

Hudson, R. (2010) *An Introduction to Word Grammar*, Cambridge: CUP

Huezo, R. (2003) *Time Linear Syntax of Spanish*, M.A. thesis, CLUE

Ickler, T. (1994) "Zur Bedeutung der sogenannten ‚Modalpartikeln'," *Sprachwissenschaft,* Vol. 19:374–404

Ide, N., K. Suderman, C. Baker, R. Passonneau, and C. Fellbaum (2010) "MASC I: Manually Annotated Sub-Corpus First Release," LDC Catalog No.: LDC2010T22, ISBN: 1-58563-569-3, Release Date: Dec 20, 2010

ISO 24610-1 (2006) "Language Resource Management – Feature Structures – Part 1: Feature Structure Representation," TC 37/SC4

Ivanova, M. (2009) *Syntax und Semantik des Bulgarischen im Rahmen von JSLIM*, M.A. thesis, CLUE

Jaakkola, H., Y. Kiyoki, and T. Tokuda (eds.) (2007) *Information Modelling and Knowledge Bases XIX*, Amsterdam: IOS Press Ohmsha

Jackendoff, R. S. (1990) *Semantic Structures*, Cambridge, MA: MIT Press

Jacobson, P. and G. K. Pullum (eds.) (1982) *The Nature of Syntactic Representation*, Dordrecht: Reidel

Jaegers, S. (2010) *Entwurf und Implementierung von Grammatikregeln für die Analyse von Komposita im Rahmen der SLIM-Sprachtheorie*, M.A. thesis, CLUE

Johnson, D. E. and P. M. Postal (1980) *Arc Pair Grammar*, Princeton: Princeton U. Press

Joshi, A. (1969) "Properties of Formal Grammars with Mixed Types of Rules and their Linguistic Relevance," *Proceedings Third International Symposium on Computational Linguistics,* Stockholm, Sweden

Jurafsky, D. and J. H. Martin (2009) *Speech and Language Processing: An Introduction to Natural Language Processing, Speech Recognition, and Computational Linguistics, 2nd ed.*, Prentice Hall

Kabashi, B. (2003) *Automatische Wortformerkennung für das Albanische*, M.A. thesis, CLUE

Kabashi, B. (2007) "Pronominal Clitics and Valency in Albanian. A Computational Linguistics Prespective and Modelling within the LAG-Framework," in: Herbst, Th., K. Götz-Votteler (Eds.) *Valency – Theoretical, Descriptive and Cognitive Issues,* pp. 339–352, Berlin, New York : Mouton de Gruyter,

Kalender, E. (2010) *Automatische JSLIM-Analyse der russischen Syntax im Rahmen der Datenbanksemantik*, M.A. thesis, CLUE

Kamp, J. A. W. (1980) "A Theory of Truth and Semantic Representation," in J.A.G. Groenendijk et al. (eds.).

Kamp, J. A. W. and U. Reyle (1993) *From Discourse to Logic*, Parts 1 and 2, Dordrecht: Kluwer

Kangassalo, H., et al. (eds.) (2003) *Information Modeling and Knowledge Bases XIV,* IOS Press Ohmsha, Amsterdam

Kapfer, J. (2010) *Inkrementelles und oberflächenkompositionales Parsen von Koordinationsellipsen*, Ph.D. thesis, CLUE

Karsten, K. B., L. N. Andriamandimbiarisoa, S. F. Fox, and C. J. Raxworthy (2009) "Sexual Selection on Body Size and Secondary Sexual Characters in 2 Closely Related, Sympatric Chameleons in Madagascar," *Behavioral Ecology,* Vol. 20.5:1079–1088

Karttunen, L. (1971) "Definite Descriptions with Crossing Coreference, A Study of the Bach-Peters Paradox," *Foundations of Language,* Vol. 7:157–182

Katz, J. and J. Fodor (eds.) (1964) *The Structure of English*, Englewood Cliffs, NJ: Prentice Hall

Kay, M. (1984) "Functional Unification Grammar: A Formalism for Machine Translation," COLING 84

Kay, M. (2004) *XPath 2.0 Programmer's Reference (Programmer to Programmer)*, Wrox Press

Kemmer, S. (2003) "Human Cognition and the Elaboration of Events: Some Universal Conceptual Categories," in M. Tomasello (ed.)

Kempson, R. and A. Cormack (1981) "Ambiguity and quantification," *Linguistics and Philosophy* 4:259–309

Kempson, R., W. Meyer-Viol, D. Gabbay (2001) *Dynamic Syntax: The Flow of Language Understanding*, Wiley-Blackwell

Kim, S. (2009) *Automatische Wortformerkennung für das Koreanische im Rahmen der LAG*, Ph.D. thesis, CLUE

Klatzky, R. L., B. MacWhinney, and M. Behrmann (eds.) (1998) *Embodiment, Ego-Space, and Action*, New York: Psychology Press

Klein, D. and Christopher D. Manning (2003) "Accurate Unlexicalized Parsing," *Proceedings of the 41st Meeting of the Association for Computational Linguistics*, pp. 423–430

Klima, E. (1964) "Negation in English," in J. Katz and J. Fodor (eds.)

Knuth, D. E. (1998) *The Art of Computer Programming. Vol. 3, 2nd ed.* Boston: Addison-Wesley

Kohonen, T. (1988) *Self-Organization and Associative Memory, 2nd ed.*, Springer

Kosche, S. (2011) *Syntax und Semantik des Französischen im Rahmen von JSLIM*, M.A. thesis, CLUE

Kripke, S. (1972) "Naming and Necessity," in D. Davidson and G. Harmann (eds.), pp. 253–355

Kučera, H. and W. N. Francis (1967) *Computational Analysis of Present-Day English*. Providence, Rhode Island: Brown U. Press

Kiyoki, Y., T. Tokuda, H. Jaakkola, X. Chen, N. Yoshida (eds.) (2008) *Information Modeling and Knowledge Bases XX*, Amsterdam: IOS Press Ohmsha

L&I'05 = Hausser, R. (2005) "Memory-Based Pattern Completion in Database Semantics," *Language and Information*, Vol. 9.1:69–92, Seoul: Korean Society for Language and Information

L&I'10 = Hausser, R. (2010) "Language Production Based on Autonomous Control - A Content-Addressable Memory for a Model of Cognition," *Language and Information*, Vol. 11:5–31, Seoul: Korean Society for Language and Information

Langacker, R. (1969) "Pronominalization and the Chain of Command," in D.A. Reibel and S. A. Shane (eds.), pp. 160–186

Lee, Kiyong (2002) "A Simple Syntax for Complex Semantics," in *Language, Information, and Computation*, Proceedings of the 16th Pacific Asia Conference, pp. 2–27

Lee, Kiyong (2004) "A Computational Treatment of Some Case-Related Problems in Korean," *Perspectives on Korean Case and Case Marking*, pp. 21–56, Seoul: Taehaksa

Leidner, J. (2000) *Linksassoziative morphologische Analyse des Englischen mit stochastischer Disambiguierung*, M.A. thesis, CLUE

Leśniewski, S. (1929) "Grundzüge eines neuen Systems der Grundlagen der Mathematik," Warsaw: *Fundamenta Mathematicae*, Vol. 14:1–81

Lieb, H. H. (1976) "Zum Verhältnis von Sprachtheorien, Grammatiktheorien und Grammatiken," in: Dieter Wunderlich (ed.). Wissenschaftstheorie der Linguistik, pp. 200–214

Lipp, S. (2010) *Automatische Wortformerkennung für das Schwedische unter Verwendung von JSLIM*, M.A. thesis, CLUE

Liu, Haitao (2009a) "Statistical Properties of Chinese Semantic Networks," *Chinese Science Bulletin*, Vol. 54(16):2781–2785

Liu, Haitao (2009b) *Dependency Grammar: From Theory to Practice*, Beijing: Science Press

Liu, Haitao (2011) "Linguistic Networks: Metaphor or Tool?" *Journal of Zhejiang University*, (Humanities and Social Science) Vol. 41.2:169–180.

MacWhinney, B. (2004) "Multiple Solutions to the Logical Problem of Language Acquisition," *Journal of Child Language,* Vol. 31:919–922. Cambridge U. Press

MacWhinney, B. (2008) "How Mental Models Encode Embodied Linguistic Perspective," in L. Klatzky et al. (eds.)

MacWhinney, B. (ed.) (1999) *The Emergence of Language from Embodiment*, Hillsdale, NJ: Lawrence Erlbaum

März, B. (2005) *Die Koordination in der Datenbanksemantik*, M.A. thesis, CLUE

Mahlow, K. (2000) *Automatische Wortformanalyse für das Spanische*, M.A. thesis, CLUE

Mahlow, K. and M. Piotrowski (eds.) (2009) *State of the Art in Computational Morphology*, Springer

McKeown, K. (1985) *Text Generation: Using Discourse Strategies and Focus Constraints to Generate Natural Language Text*, Cambridge: CUP

Mehlhaff, J.(2007) *Funktor-Argument-Struktur und Koordination im Deutschen – eine Implementation in JSLIM*, M.A. thesis, CLUE

Mei, Hua (2007) *Automatische Syntax und Semantikanalyse des Chinesischen in Rahmen von JSLIM*, M.A. thesis, CLUE

Michalski, R. S. and R. E. Stepp (1983). "Learning from Observation: Conceptual Clustering," in Michalski, R. S.; Carbonell, J. G.; Mitchell, T. M. (Eds.) *Machine Learning: An Artificial Intelligence Approach*, pp. 331–363, Palo Alto, CA: Tioga

Miller, G. (1995) "Wordnet: A Lexical Database," *Communications of the ACM*, Vol. 38.11:39–41

Minsky, M. (1975) "A Framework for Representing Knowledge," in P. Winston (ed.), pp. 211–277

Mitchell, T. (1997) *Machine Learning*, McGraw-Hill

Mithun, M. (1999) *The Languages of Native North America*. Cambridge U. Press

Mohammadzadeh, A., S. Soltan, and M. Rahgozar (2006), "Template Guided Association Rule Mining from XML Documents," in *Proceedings of the 15th International Conference on World Wide Web*, pp. 963–964, New York: ACM Press

Montague, R. (1974) *Formal Philosophy*, New Haven: Yale U. Press

Muñoz-Avila, H., U. Jaidee , D. W. Aha, and E. Carter (2010) "Goal-Driven Autonomy with Case-Based Reasoning," in Proceedings of ICCBR2010, pp. 228–241

Naphade, M. R. and J. R. Smith (2009) "Computer Program Product and System for Autonomous Classification," Patent Application #:20090037358 - Class: 706 46 (USPTO)

Nichols, J. (1992) *Linguistic Diversity in Space and Time*, The U. of Chicago Press

Niedobijczuk, K.(2010) *Implementierung eines automatischen Wortformerkennungssystems des Polnischen in JSLIM*, M.A. thesis, CLUE

NLC'06 = Hausser, R. (2006) *A Computational Model of Natural Language Commmunication: Interpretation, Inference, and Production in Database Semantics*, Springer

Pakula, A. J. (1976) *All the President's Men*, Warner Brothers

Pandea, P. (2010) *Implementierung eines automatischen Wortformerkennungssystems im Rumänischen mit JSLIM*, M.A. thesis, CLUE

Peirce, C. S. (1933) *The Simplest Mathematics*, Collected Papers, Vol.4, C. Hartshorne and P. Weiss (eds.), Cambridge, MA: Harvard U. Press

Pepiuk, S.   (2010) *Implementierung eines automatischen Wortformerkennungssystems für das Französische in JSLIM*, M.A. thesis, CLUE

Perlmutter, D. M.   (1980) "Relational grammar," In E. A. Moravcsik and J. R. Wirth (Eds.), *Syntax and Semantics: Current Approaches to Syntax*, Vol. 13:195–229. New York: Academic Press

Peters, S. and Ritchie, R.   (1973) "On the Generative Power of Transformational Grammar," *Information and Control*, Vol. 18:483–501

Petrov, S. and D. Klein   (2007) "Improved Inference for Unlexicalized Parsing" in proceedings of HLT-NAACL

Pflug, I. J.   (1967/1999) *Microbiology and Engineering of Sterilization Processes, 10th edition*, Environmental Sterilization Laboratory, Minneapolis, MN

Proisl, T.   (2008) *Integration von Valenzdaten in die grammatische Analyse unter Verwendung des Valency Dictionary of English*, M.A. thesis, CLUE

Proust, M.   (1913) *Du côté de chez Swann*, ed. by Jean-Yves Tadie et al., Bibliothèque de la Pléiade, Paris: Gallimard,1987–1989

Pullum, G. K.   (1991) *The Great Eskimo Vocabulary Hoax: And Other Irreverent Essays on the Study of Language*, The U. of Chicago Press

Pustejovsky, J.   (1995) *The Generative Lexicon*, Cambridge, MA: MIT Press

Quillian, M.   (1968) "Semantic Memory," in M. Minsky (ed.), *Semantic Information Processing*, pp. 227–270, Cambridge, MA: MIT Press

Quine, W. v. O.   (1960) *Word and Object*. Cambridge, MA: MIT Press

Quirk R., S. Greenbaum, G. Leech, and J. Svartvik (1985) *A Comprehensive Grammar of the English Language*, London New York: Longman

Reibel, D. A. and S.A. Shane (eds.) (1969) *Modern Studies of English*, Englewood Cliffs, New Jersey: Prentice Hall

Reihl, S.   (2010) *Inkrementelle und oberflächenkompositionale Analyse von Derivaten im Deutschen*, M.A. thesis, CLUE

Reis, M. and W. Sternefeld   (2004) "Review of S. Wurmbrand: Infinitives. Restructuring and Clause Structure," *Linguistics*, Vol. 42:469–508

Rijkhoff, J.   (2002) *The Noun Phrase*. Oxford U. Press

Rosch, E.   (1999) "Reclaiming Concepts", Journal of Consciousness Studies, Vol.6.11–12:61–77

Ross, J. R. (1969) "On the cyclic nature of English pronominalization," in D.A. Reibel and S. A. Schane (eds.), pp. 187–200

Roy, D. (2003) "Grounded spoken language acquisition: experiments in word learning," *IEEE Transactions on Multimedia*, Vol. 5.2:197–209

Roy, D. (2005) "Grounding words in perception and action: computational insights," *TRENDS in Cognitive Sciences*, Vol. 9.8:389–396

Roy, D. (2008) "A Mechanistic Model of Three Facets of Meaning. Symbols, Embodiment, and Meaning," in de Vega, Glenberg, and Graesser (eds.) *Symbols and Embodiment: Debates on Meaning and Cognition*, Oxford U. Press

Sabeva, S. (2010) *Erstellung eines linksassoziativen morphologischen Analysesystems für das Bulgarische*, Ph.D. thesis, CLUE

Sag, I. and T. Wasow (2011) "Performance-Compatible Competence Grammar," in R. Borsley and K. Borjars (eds.) *Non-Transformational Syntax*, Wiley-Blackwell

Sasse, H. J. (1993) "Das Nomen – eine universelle Kategorie?" in *Sprachtypologie und Universalienforschung*, Vol. 46,3:187–221

Saussure, F. de (1916/1972) *Cours de linguistique générale*, Édition critique préparée par Tullio de Mauro, Paris: Éditions Payot

SCG'84 = Hausser, R. (1984) *Surface Compositional Grammar*, Munich: Wilhelm Fink Verlag

Schank, R. (1982) *Dynamic Memory: A Theory of Learning in Computers and People,* New York: Cambridge U. Press

Schegloff, E. (2007) *Sequence Organization in Interaction,* New York: Cambridge U. Press

Sells, P. (1985) *Lectures on Contemporary Syntactic Theory: an introduction to government-binding theory, generalized phrase structure grammar, and lexical-functional grammar*, Stanford: CSLI Lecture Notes Number 3

Schnelle, H. (1988) "Ansätze zur prozessualen Linguistik," in: Schnelle, H. and G. Rickheit (eds.), *Sprache in Mensch und Computer*, pp. 137–190

Schnelle, H. and G. Rickheit (eds.) (1988) *Sprache in Mensch und Computer*, Opladen: Westdeutscher Verlag.

Shopen, T., (ed.) (1985)*Language Typology and Syntactic Description III: Grammatical Categories and the Lexicon*, CUP

Sinclair, J. (1991) *Corpus Concordance Collocation*, Oxford: Oxford U. Press

Söllch, G. (2009) *Syntaktisch-semantische Analyse des Tagalog im Rahmen der Datenbanksemantik*, M.A. thesis, CLUE

Solé, R. V., B. Corominas-Murtra, S. Valverde, and L. Steels (2010) "Language networks: their structure, function, and evolution," *Complexity*, Vol. 15.6:20–26

Sorensen, M. N., (1997) "Infinitives (and more)," http://faculty.washington.edu/marynell/grammar/infini.html

Sowa, J. (1987/1992) "Semantic networks," *Encyclopedia of Artificial Intelligence*, edited by S. C. Shapiro; revised and extended for the second edition, 1992. New York: Wiley
Available at http://www.jfsowa.com/pubs/semnet.htm

Steels, L. (1999) *The Talking Heads Experiment*. Antwerp: limited pre-edition for the Laboratorium exhibition

Stefanskaia, I. (2005) *Regelbasierte Wortformerkennung für das Deutsche als Grundlage eines linksassoziativen Sprachmodells für einen automatischen Spracherkenner mit Prototyping einer Sprachsteuerungskomponente zur Steuerung dezidierter Arbeitsabläufe in einem radiologischen Informationssystem*, M.A. thesis, CLUE

Talmy, L. (1985) "Lexicalization Patterns: Semantic Structure in Lexical Forms," in T. Shopen (ed.), pp. 57–149

Tarski, A. (1935) "Der Wahrheitsbegriff in den Formalisierten Sprachen," *Studia Philosophica*, Vol. I:262–405

TCS'92 = Hausser, R. (1992) "Complexity in Left-Associative Grammar," *Theoretical Computer Science*, Vol. 106.2:283–308

Tesnière, L. (1959) *Éléments de syntaxe structurale*, Editions Klincksieck, Paris

Tilden, M. W. and B. Hasslacher, (1996) "Living Machines," Los Alamos National Laboratory, Los Alamos, NM 87545, USA

Tittel, A. (2008) *Optimierung der deutschen JSLIM-Morphologie*, M.A. thesis, CLUE

Tomasello, M. (2003) "Introduction: Some Surprises to Psychologists," in M. Tomasello (ed.)

Tomasello, M. (ed.) (2003) *The New Psychology of Language, Vol. II*, Mahaw, N.J.: Lawrence Erlbaum

Twiggs, M. (2005) *Eine Behandlung des Passivs im Rahmen der Daten-banksemantik,* M.A. thesis, CLUE

Vogel, P. and B. Comrie (eds.) (2000) *Approaches to the Typology of Word Classes,* Berlin: Mouton de Gruyter

Vorontsova, L. (2007) *Implementierung eines automatischen Wortformerken-nungssystems für das Russische in JSLIM,* M.A. thesis, CLUE

Webb, B. and J. Wessnitzer (2009) "Perception for Action in Insects," in Arena, P. and L. Patanè (eds.)

Weber, C. (2007) *Implementierung eines automatischen Wortformerkennungs-systems für das Italienische mit dem Programm JSLIM,* M.A. thesis, CLUE

Weber, C. and J. Handl (2010) "A Base-form Lexicon of Content Words for Correct Word Segmentation and Syntactic-Semantic Annotation", in M. Pinkal, J. Rehbein, S. Schulte im Walde, and A. Storrer (eds.) *Semantic Approaches in Natural Language Processing. Proceedings of the Conference on Natural Language Processing (KONVENS) 2010,* pp. 175–179, Saar-brücken: universaar

Weiss, M.A. (2005) *Data Structures and Problem Solving Using Java. 3rd ed.,* Upper Saddle River, NJ: Pearson Addison-Wesley

Weizenbaum, J. (1965) "ELIZA – A Computer Program for the Study of Natural Language Communications between Man and Machine," *Commun. ACM,* Vol. 9.1 (11)

Whorf, B. (1964) *Language, Thought, and Reality: Selected Writings of Benjamin Lee Whorf,* J. B. Carroll (ed.), MIT Press

Wiener, N. (1948) *Cybernetics: Or the Control and Communication in the Animal and the Machine,* Cambridge, MA: MIT Press

Wierzbicka, A. (1991) *Cross-Cultural Pragmatics: The Semantics of Human Interaction.* Berlin, New York: Mouton de Gruyter

Wilson, E. O. (1998) *Consilience: The Unity of Knowledge,* New York: Alfred Knopf

Winston, P. (ed.) (1975) *The Psychology of Computer Vision,* New York: McGraw-Hill

Wu, Z. and M. Palmer (1994) "Verb Semantics and Lexical Selection," Proceedings of the 32nd annual meeting of the ACL, pp. 133–138

Wunderlich, D. (ed.) (1976) *Wissenschaftstheorie der Linguistik,* Kronberg: Athenäum (= Athenäum Taschenbücher Sprachwissenschaft)

Wurmbrand, S. (2001) *Infinitives. Restructuring and Clause Structure*. Studies in Generative Grammar 55. Berlin: Mouton de Gruyter

Yovits, M. C. (ed.) (1992) *Advances in Computer Science, 2nd ed.*, Academic Press

Zholkovskij, A. and Mel'chuk, A. (1965) "O vozmozhnom metode i instrumentax semanticheskogo sinteza." (On a possible method and instruments for semantic synthesis.) Nauchno-texnicheskaja informacija. Retrieved May, 19, 2001, from the World Wide Web: http://www.neuvel.net/meaningtext.htm

Zimmerman, D. (1991) "Structure-in-Action: an Introduction," in D. Boden and D. Zimmerman (eds.)

Zipf, G. K. (1932) *Selected Studies of the Principle of Relative Frequency in Language*, Cambridge, MA.: Harvard U. Press

Zipf, G. K. (1935) *The Psycho-Biology of Language*, Cambridge, MA: MIT Press

Zipf, G. K. (1949) *Human Behavior and the Principle of Least Effort*, Cambridge, MA: Addison-Wesley

Zue, V., R. Cole, and W. Ward (1995) "Speech Recognition," in R. Cole (ed.) 1997

Zweig, E. (2008) *Dependent Plurals and Plural Meaning*, Ph.D. dissertation, the University of York

Name Index

Subject Index

Subject Index

now front, 56, 58, 67, 78, 89, 100, 105, 195, 246

object control, 167
object gapping, 187
ontology, 11, 225, 246
optical illusion, 153
Optimality Theory, 145

part of speech (PoS), 2, 45, 52, 114, 128, 140
particles, 207
passive voice, 174
pattern matching, 57, 60
PCFG, 13
personnel, 13, 239
perspective conversion, 210
Phrase Structure Grammar, 11, 108, 121, 154, 169
placeholder value, 98, 115, 162
pointer, 77
politeness, 118, 193, 199, 207
predicate calculus, 102, 224, 243, 244
prepositional object, 138
prepositional phrase, 126, 138, 155
proplet, 6
propositional calculus, 95, 102, 243

quantifier, 102, 106, 224, 244
query pattern, 57
question, 57, 199, 204

realism (mathematical), 23, 64
recognition by components, 99
recursion, 154
recursive ambiguity, 9, 156
reference, 58, 61, 63–65
reference corpus, 31, 165, 238
reflex, 100
RegEx, 158

register adaptation, 237
relative clause, 178
renewable language data, 238, 251
request, 206
RMD corpus, 31, 161, 165, 169, 238
robustness, 28, 249

saccades, 150
schema, 37, 58, 106
schnurzpiep(egal), 52
search space reduction, 30, 77, 82, 162
sediment, 62
selectional constellations, 160, 162, 164, 165, 169
self-organization, 76
semantic hierarchy, 10, 108, 144
sentential argument, 218
sentential mood, 199, 208
Sequential Inferencing Principle (SIP), 76
service channel, 29, 236
sign theory, 25, 47, 48, 60, 127
situation awareness, 193
Situation Semantics, 23
SLIM theory of language, 1, 4, 10, 64
smart solution, 28
social animals, 73
soft, 195
speak mode, 40
speech recognition, 26, 249
spreading activation, 82
SQL, 66
standardization, 212, 239
statistics, 13, 30, 239
subactivation, 81, 162
subclass relation, 108, 144
subject control, 160, 168